COLLINS GEM

MATHEMATICS

BASIC FACTS

C. Jones BSc & P. Clamp BSc
revised by R. Browne BA

HarperCollins*Publishers*

HarperCollins Publishers
PO Box, Glasgow G4 0NB, Scotland

First published 1982
Revised edition 1988
Third edition 1991

Reprint 10 9 8 7 6 5

ISBN 0 00 470174 7

Printed in Great Britain by
HarperCollins Manufacturing, Glasgow

Introduction

Collins Gem *Basic Facts* is a series of illustrated GEM dictionaries in important school subjects. This new edition has been extensively revised and updated to widen the coverage of the subject and to reflect recent changes in the way it is taught in the classroom.

Bold words in an entry identify key terms which are explained in greater detail in entries of their own; important terms which do not have separate entries are shown in *italic* and are explained in the entry in which they occur.

Other titles in the series include:

Gem *Computers Basic Facts*
Gem *Chemistry Basic Facts*
Gem *Physics Basic Facts*
Gem *Science Basic Facts*
Gem *Biology Basic Facts*
Gem *History Basic Facts*
Gem *Geography Basic Facts*
Gem *Technology Basic Facts*
Gem *Business Studies Basic Facts*

Symbols in Mathematics

Symbol	Meaning	Example
$=$	equals	$3 = 2+1$
$+$	add	$3+4 = 7$
$-$	subtract, negative	$3-4 = -1$
$\times$	multiply	$3 \times 4 = 12$
$\div$	divide	$3 \div 4 = 0.75$
$\sqrt{}$	square root	$\sqrt{9} = \pm 3$
n	power	$3^2 = 9$
$\angle$	angle	$\angle ABC$
$\triangle$	triangle	$\triangle ABC$
$<$	less than	$1 < 4$
$\leqslant$	less than, or equal to	$x \leqslant y$
$>$	greater than	$6 > 4$
$\geqslant$	greater than, or equal to	$y \geqslant x$
$\neq$	not equal to	$4.9 \neq 49$
$\simeq$	approximately equal to	$3.9 \simeq 4$
$\propto$	proportional to	$x \propto y$
$\{\ \}$	denotes a set	$\{1, 2, 3, \dots\}$
$\in$	is a member of	$1 \in \{1, 2, 3\}$
$\subset$	is a subset of	$A \subset B$
$\supset$	includes the subset	$B \supset A$
$\mathscr{E}$	universal set	$A \subset \mathscr{E}$
$\varnothing$	empty set	$\varnothing = \{\ \}$
$\cap$	intersection	$A \cap B = \varnothing$
$\cup$	union	$A \cup B = \mathscr{E}$
$\Rightarrow$	implies	$x = 2 \Rightarrow x^2 = 4$
$\therefore$	therefore	
$\equiv$	congruent to	

A

abacus A frame containing **parallel** rods carrying beads, used for computation. In the Chinese abacus shown below, the beads above the bar are worth five times the beads below.

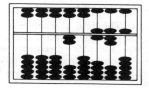

abacus The number shown at the bar is 20 675.

abscissa The *x*-coordinate, or distance from the *y*-axis, of a **point** referred to a system of **Cartesian coordinates**.

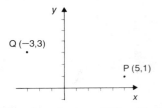

abscissa   The abscissa of Q is −3 and of P is 5.

absolute value   The numerical **value** of a **real number** without regard for **sign**. The absolute value of the real number x (sometimes called **modulus** or **mod** x) is written $|x|$. For example, $|3.6|=3.6$ and $|-9.1|=9.1$.

acceleration   The rate of change of **velocity** with time. For example, if a body falls from rest with an acceleration of 9.8 metres per second per second, its velocity at intervals of a second, from 0 to 3 seconds, would be as shown in the table below:

Time (s)	Velocity (m s)
0	0
1	9.8
2	19.6
3	29.4

acceleration

If **v** is a **vector** representing the velocity of a particle, then the acceleration of the particle is a vector **a** obtained by differentiating **v** with respect to time

$$\mathbf{a} = \frac{d\mathbf{v}}{dt}$$

The acceleration of a particle is thus also represented by the **gradient** of the particle's velocity–time graph. See **differentiation**.

acute angle An **angle** that measures between 0 and 90 **degrees**. The opposite of an acute angle is an **obtuse angle**.

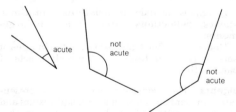

acute Examples of angles that are and are not acute.

addition The process of adding numbers together which is one of the basic **operations** of **arithmetic**. The **sum** of two numbers is determined by the operation addition, which is related to the process of accumulation, for example, $3+2=5$.

In abstract **algebra** used to denote certain operations applied to various **sets**, for example, in the set of 2×2 real **matrices**, addition is defined by:

$$\begin{pmatrix} a & b \\ c & d \end{pmatrix} + \begin{pmatrix} e & f \\ g & h \end{pmatrix} = \begin{pmatrix} a+e & b+f \\ c+d & d+h \end{pmatrix}$$

affine transformation A **transformation** of the **plane** that can be represented in the form

$$\begin{pmatrix} x \\ y \end{pmatrix} \rightarrow \begin{pmatrix} a & b \\ c & d \end{pmatrix} \begin{pmatrix} x \\ y \end{pmatrix} + \begin{pmatrix} e \\ f \end{pmatrix}$$

Examples include **translations**, **rotations**, **shears**, **reflections**, stretches and **enlargements**.

Under any affine transformation the **images** of sets of **parallel lines** are themselves sets of parallel lines.

algebra The generalization, and representation in symbolic form, of significant results and patterns in **arithmetic** and other areas of mathematics. For example,

$$(4+3)(4-3)=4^2-3^2$$

and

$$(9+5)(9-5)=9^2-5^2$$

are both examples of the algebraic statement

$$(x+y)(x-y)=x^2-y^2$$

algorithm A standard process designed to solve a particular set of problems. The familiar algorithm for the **addition** of **fractions** of the form

$$\frac{w}{x}+\frac{y}{z}$$

is as follows:

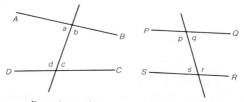

| find d
the LCM
of x
and z | find
$q_1 = \frac{d}{x}$ | find
$n_1 = q_1 \times w$ | find
$q_2 = \frac{d}{z}$ | find
$n_2 = q_2 \times y$ | find
$n = n_1 + n_2$ | result
$= \frac{n}{d}$ |

alternate angles A pair of **angles** that are on opposite sides of a **transversal** cutting two lines, with each angle having one of the lines for one of its sides.

alternate angles *a* and *c*, *b* and *d*, *p* and *r*, and *q* and *s* are all alternate angles.

In the diagram a and c are an alternate pair of angles, as are b and d. Lines PQ and SR are **parallel**, and in this case the alternate pairs of angles are equal, $p = r$ and $q = s$.

altitude In a **triangle**, a **perpendicular** distance from a **vertex** of the triangle to the opposite side of the triangle, called the **base**.

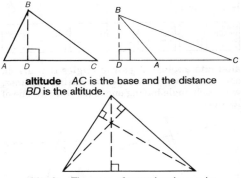

altitude AC is the base and the distance BD is the altitude.

altitude The area of any triangle can be calculated by $\frac{1}{2} \times$ base length $\times$ altitude.

Note that the three altitudes of any triangle pass through a common point as shown in the figure.

amplitude A measure of the maximum **magnitude** of the **displacement** from the **mean** or base position in an oscillating motion.

For the periodic function $f(x) = 3 \sin x$, the amplitude is 3 (see **period**).

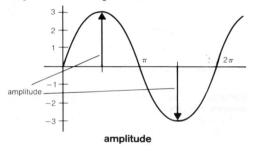

amplitude

analysis The branch of mathematics concerned with rigorous proofs of propositions, mainly in the area of **calculus**.

angle **1.** A measure of the space between two straight lines that extend from a common **point**. **2.** A measure of the **rotation** (about the point of **intersection** of two lines) required to map one line onto the other.

The two most common measures of angle are **degree** and **radian** measure. See **acute angle**, **obtuse angle**.

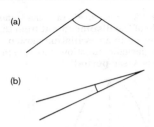

(a)

(b)

angle Large (a) and small (b) angles.

annulus A **region** of a **plane** bounded by two **concentric circles**.

 If the circles are of **radii** r and R, with $r<R$, then the **area** of the annulus is:

$$\pi R^2 - \pi r^2 \quad \text{or} \quad \pi(R+r)(R-r)$$

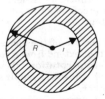

annulus The shaded region is an annulus.

antilogarithm   The **inverse** of a **logarithm function**. It is used particularly when base 10 logarithms are employed specifically for calculating purposes. For example,

$$\log_{10} 100 = 2 \qquad \text{antilog}_{10} 2 = 100$$

apex   The highest point of a solid or **plane** figure relative to a base plane or line.

apex   apex   apex

base   base   base

apex   The apexes of some solid and plane figures.

approximation   An inexact result, which is accurate enough for some specific purpose. For example, to the nearest tenth, the **square root** of 2 is 1.4. To the nearest hundredth it is 1.41.

The study of processes for approximating various forms in mathematics is an important branch of the subject. For example, *Taylor's Theorem* provides a way of approximating to certain **functions** by **polynomials**.

Newton's method gives an **iterative** way of obtaining approximations to the **roots** of certain **equations**.

arc A part of a **curve**. In particular for the **circle**, the points A and B define both **minor** and **major** arcs of the circle.

In the diagram, for the minor arc AB:

Arc length $= 2\pi r \times \theta / 360$, if θ is in **degrees**

or

Arc length $= r\theta$, if θ is in **radians**.

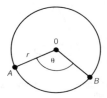

arc A circle showing its minor and major arcs.

area The measure of the size of a surface. The areas of certain simple figures can be easily calculated.

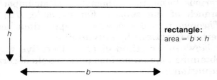

rectangle:
area $= b \times h$

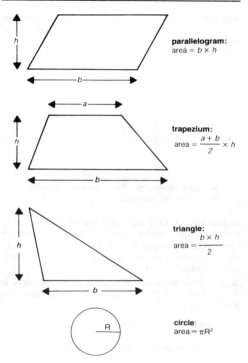

parallelogram:
area = $b \times h$

trapezium:
area = $\dfrac{a + b}{2} \times h$

triangle:
area = $\dfrac{b \times h}{2}$

circle:
area = πR^2

Argand diagram A rectangular **Cartesian** grid used for diagrammatic representations of **complex numbers**.

The horizontal axis is called the *real axis*, and the vertical axis is called the *imaginary axis*.

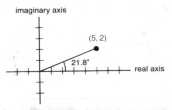

Argand diagram The complex number $5+2i$ is represented by the point (5, 2).

argument 1. Of a **complex number** the **angle** between the position **vector** of the number on the **Argand diagram**, and the positive real axis. In the diagram above, the argument of the complex number $5+2i$ is 21.8°.

2. The **independent variable** of a **function** is sometimes called the argument of the function. For example:

x is the argument of the function $y=3x^2+6x-7$

f is the argument of the function $C=\dfrac{(f-32)\times 5}{9}$

arithmetic The study of numbers, particularly with regard to simple **operations**: **addition**, **subtraction**, **division** and **multiplication**, and their application to solutions of problems.

arithmetic mean Commonly referred to as the **average** of a given set of numbers. The arithmetic mean of n numbers a_1, a_2, ... a_n is calculated by

$$(a_1 + a_2 + a_3 + \ldots a_n) \div n.$$

For example, the arithmetic mean of 5, 7, 1, 8, 4 is

$$(5 + 7 + 1 + 8 + 4) \div 5 = 5.$$

arithmetic progression A **sequence** in which there is **common difference** between any member of the sequence and its successor.

For example, 3, 7, 11, 15, 19, 23 is an arithmetic progression with common difference 4.

associative Any **binary operation** $*$ which has the property $a*(b*c) = (a*b)*c$ for all members of a, b and c of a given set.

Multiplication of **real numbers** is associative, for example:

$$8 \times (5 \times 2) = (8 \times 5) \times 2.$$

Division is not associative, for example:

$$80 \div (4 \div 2) \neq (80 \div 4) \div 2.$$

asymptote A line approached, but never reached, by a **curve**.

The curve on the **graph** can be drawn as close to $y=3$ as desired by taking sufficiently large values of x.

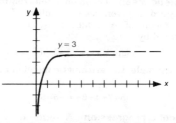

asymptote $y=3-1/x$ has the lines $y=3$ and $x=0$ as asymptotes.

average A single number that represents or typifies a collection of numbers. **Mode**, **mean** and **median** are three commonly used averages.

axiom A principle, taken to be self-evident, and not requiring proof.

'Things equal to the same thing are equal to each other' is one of the axioms attributed to Euclid (see **Euclidean geometry**).

axis (plural, axes) A significant line which a **graph**, **plane** shape or solid can be referenced to.

For example, **coordinate** axes, axes of **symmetry**, etc.

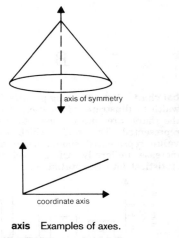

axis Examples of axes.

B

bar chart A **graph** using **parallel** bars of equal width to illustrate information. The lengths of the bars are proportional to the quantity represented. The **scale** which indicates the **value** represented should begin at **zero** and increase uniformly. Often used to illustrate statistical data. See **statistics**.

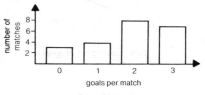

bar chart

bar-line graph A method of displaying information very similar to a **bar chart**, but using bar-lines instead of bars.

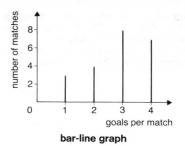

bar-line graph

base   1. Number base is the method of grouping used in a number system that relies on place **value**. In the **decimal** number system, grouping and place value is based on 10. For example,

in base 10, 423 means $3 + (2 \times 10) + (4 \times 100)$;

in base 5, 423 would mean $3 + (2 \times 5) + (4 \times 25)$.

2. In expressions like 4^5, 4 is called the base whilst 5 is called the **exponent**. This use of the term base is closely linked to its use in the context of **logarithms**.

In base 10 logarithms log $47 = 1.6721$, this means that $10^{1.6721} = 47$.

3. In **geometry** the lower side of a **plane** figure, such as a **triangle**, or of a solid, such as a **pyramid**.

bearing A navigational and surveying term describing direction. The bearing of a point *A* from an observer *O* is the **angle** between the line *OA* and the north line through *O*, measured in a **clockwise** direction from the north line.

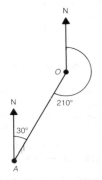

bearing The bearing of *A* from *O* is 210°. The three figure bearing of *O* from *A* is 030°.

billion A thousand million, that is, 10^9. In the USA the term has always had this meaning, while in Britain it used to mean a million million, that is, 10^{12}. The meaning 10^9 is now common to both countries.

bimodal A term used to describe **distributions** of data that show two **modes** or peaks in **frequency**.

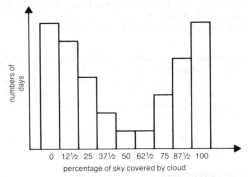

numbers of days

0 12½ 25 37½ 50 62½ 75 87½ 100
percentage of sky covered by cloud

bimodal The distribution is bimodal with modes 0 and 100.

binary Denoting or based on two.
1. A binary number system is based on grouping in twos. Binary notation requires the use of just two symbols 0 and 1, and its column headings or place values are **powers** of 2:

... 64 32 16 8 4 2 1

Therefore

29 is:			1	1	1	0	1	in binary,
92 is:	1	0	1	1	1	0	0	in binary.

2. A binary **operation** is a rule for combining two **members** of a **set** to produce a third member of the set. The four basic operations of **arithmetic** — **addition**, **subtraction**, **multiplication** and **division** — are examples of binary operations on the set of **real numbers**.

binomial   An algebraic expression containing two variables, for example, $7x+4y$.

The numbers in successive rows correspond to the coefficients of the terms when $(x+y)^n$ is expanded in **powers** of x and y for various values of n. For example:

$$(x+y)^5 = x^5+5x^4y+10x^3y^2+10x^2y^3+5xy^4+y^5$$

See **polynomial**.

binomial   The numbers in the rows of **Pascal's triangle** are called binomial **coefficients**.

bisect To cut in half. The term is often used in a geometrical context.

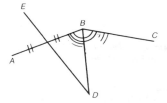

> **bisect** The line *BD* bisects ∠ *ABC*. The line *DE* bisects the line *AB*.

block graph A diagram indicating the amounts accorded to various quantities in a particular context. For instance, it might show how many children in a class considered each of several colours to be their favourite colour.

A block **graph** should have a baseline, equal-sized blocks, and some way of reading off the total number of blocks in each column.

boundary The **curves** which define the outer **edges** of a figure.

The boundary of a *polygon* consists of the edges; of a **circle**, its **circumference**.

(a)   (b)

boundary The boundaries of (a) a polygon and (b) a circle.

brackets A pair of symbols used to indicate the order in which operations are to be carried out. For example:

$$7+(2\times3) = 7+6=13 \quad (7+2)\times3=9\times3=27$$

Elementary **algebra** is often concerned with the introduction or removal of brackets in cases where one operation distributes over another. For example:

$$6x^2y+9xy^2 = 3xy(2x+3y)$$

C

calculate To perform a mathematical process, often numerical, to obtain some desired result.

calculus An important branch of mathematics concerned with the study of the behaviour of **functions**.

(a) **Differential** calculus is concerned with the variation of functions, with maximum and minimum **values, gradients**, approximations to functions, etc.

(b) **Integral** calculus is concerned with calculations of **areas** and **volumes**, problems of summation, etc.

The techniques of calculus were developed independently by Newton and Leibniz in the 17th century. Calculus has important applications in many areas of the natural, physical and social sciences.

cancellation The process of producing, from a

given **fraction**, an **equivalent** fraction, by dividing **numerator** and **denominator** by a common **factor**. For example:

$$\frac{15}{20} = \frac{(3 \times 5) \div 5}{(4 \times 5) \div 5} = \frac{3}{4}$$

See **equivalent fractions**.

cardinal number   That aspect of a number that expresses how many **elements** are in a **set**. The cardinal number 3 is the property shared by all sets containing three elements.

cardioid   The **locus** or *path* of a **point** on the **circumference** of a **circle**, which rolls on a fixed circle of equal **radius**.

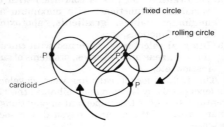

cardioid   The path of one circle rolling round another of the same radius.

Cartesian coordinates A method of locating the position of a **point** on a **plane** or in space.

In the case of a plane the distances of the point from two perpendicular axes, are given as an ordered pair of **real numbers** (x, y).

In space, the distances of the point from three mutually perpendicular planes, which in turn form three mutually perpendicular axes, as they intersect in pairs. The distances are given as an ordered triple of real numbers (x, y, z).

(Named after the French scientist and mathematician René Descartes, 1596–1650, who devised this system.)

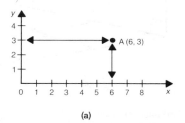

(a)

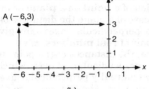

(b)

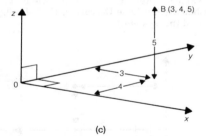

(c)

Cartesian coordinates The coordinates of the point A are (a) (6, 3) and (b) (−6, 3). The coordinates of point B (c) are (3, 4, 5).

catenary The shape of the curve assumed by a heavy, flexible cable or rope, when suspended from two points. The **graph** of $y=\cosh x$ is a catenary.

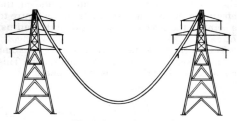

catenary The shape formed by cable suspended from two points.

centigrade The **unit** of temperature in which the freezing and boiling temperatures of water are assigned 0 and 100 degrees respectively. 37 degrees centigrade is written 37°C. See **Fahrenheit**.

centimetre One-hundredth of a **metre**.

centimetre A penny is 2 cm in **diameter**.

centre A point which is equidistant from the **boundary** of a figure. The centre of a **circle** is the point equidistant from all points on the **circumference**. The distance from the centre to a point on the circumference is called the **radius**, *R*.

The centre of a **sphere** is the point equidistant from all points on the surface of the sphere. This distance is called the radius of the sphere.

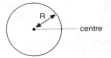

centre The radius is the distance from the centre to a point on the circumference.

characteristic When using common, **base 10 logarithms**, the characteristic is the **integer** part of the logarithm when written in the conventional way. The characteristic conveys information regarding the position of the decimal point in the original number. For example:

$$0.041 \quad = 10^{-2} \times 4.1$$
$$\log(0.041) = \log(10^{-2} \times 4.1)$$
$$= \log(10^{-2} + \log(4.1)$$
$$= -2 + 0.6128$$
$$\text{written } \bar{2}.6128$$

$\bar{2}$ is the characteristic of the logarithm.

chord A line joining two points on a curve.

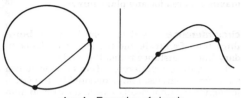

chord Examples of chords.

circle A **plane** curve formed by the **set** of all points at a given fixed distance from a fixed point. The fixed point is called the **centre**, and the distance the **radius**.

A circle of radius R has area πR^2 and circumference $2\pi R$.

Circles have line and rotational **symmetry** of **infinite** order about the centre.

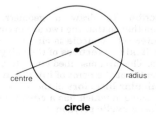

centre   radius

circle

For a fixed **perimeter**, a circle encloses the maximum area for any plane **curve**.

circumference   Used to mean both the **boundary** line of a **circle** and the length of the boundary line, or distance round the circle.

The ratio circumference/**diameter** is a constant for all circles, and is denoted by the Greek letter π (**pi**). Numerically π is an **irrational** number of approximate **value** 3.14 (to three significant figures). The circumference of a circle may therefore be calculated using the formulae:

circumference = π×diameter.

For example, for a circle of diameter 6 cm:

circumference ≃ 3.14×6 cm(≃18.8 cm).

circumscribe   To draw a geometric figure around another so that the two are in contact but do not intersect. A **circle** is circumscribed to a **polygon** if all the **vertices** of the polygon lie on the circle. The circumscribed circle to a triangle has its centre at the point of **intersection** of the **perpendicular bisectors** of the sides of the triangle. A polygon which has a circumscribed circle is called a **cyclic** polygon.

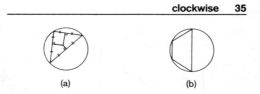

circumscribe A circle circumscribed to (a) a triangle and (b) a polygon.

class interval A grouping of statistical **data** to enable the data to be represented and interpreted in a simpler way. For example, the following 50 examination percentages

> 29, 33, 49, 71, 16, 53, 62, 81, 31, 26,
> 39, 46, 52, 42, 31, 26, 11, 62, 77, 21,
> 69, 26, 60, 21, 44, 19, 79, 50, 39, 54,
> 43, 48, 55, 27, 12, 42, 70, 74, 40, 70,
> 29, 33, 60, 53, 48, 31, 15, 88, 45, 62,

could be grouped into these intervals:

Mark	1–25	26–50	51–75	76–100
Frequency	7	24	15	4

clockwise An indication of direction of **rotation** in a **plane**. A clockwise rotation has the same sense as the rotation of the hands of a clock.

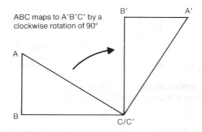

ABC maps to A'B'C' by a clockwise rotation of 90°

clockwise ABC maps to A'B'C' by a clockwise rotation of 90°.

closed   1. A **set** is described as closed under an **operation** when the result of combining any two members of the set, using the operation, always results in a member of the original set.

For example, the set of **natural** numbers {1, 2, 3, 4, ...} is closed under the operations of **addition** and **multiplication**, but is not closed under **division** or **subtraction**:

$3 \div 2 = 1.5$, which is not a natural number
$1 - 4 = -3$, which is not a natural number

2. A closed curve is one with no end points. See **simple closed curve**.

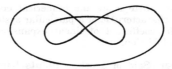

closed   A closed curve.

3. A closed **interval** on the **real number** line is a set of all numbers x, defined by **inequalities** of the type $a \leqslant x \leqslant b$.

The closed interval $\{x: 1 \leqslant x \leqslant 3\}$ is the set of all real numbers between, and including, the 'end points' 1 and 3. It is denoted by [1, 3].

closed   A closed interval $\{x: 1 \leqslant x \leqslant 3\}$.

coefficient   1. In simple **algebra**, the numerical part of a term. For example, in $4x^2y + 3xy^2 - 12x + 9y$:

> the coefficient of x^2y is 4
> the coefficient of xy^2 is 3
> the coefficient of x is -12
> the coefficient of y is 9

2. In science and engineering the term is often

used to denote a specific numerical **constant** that is characteristic of a particular system, for example, coefficient of linear expansion, coefficient of friction, etc.

collinear   Said of a set of **points** lying in a straight line.

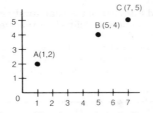

collinear   The three points A, B and C are collinear.

Notice in the **graph** that the **line segments** AB and BC are related: AB=2BC. Notice also that there is a connection between the position of A, B and C: OB=$\frac{1}{3}$OA+$\frac{2}{3}$OC and B divides the line segment AC in the **ratio** 2:1.

column   A vertical array of **elements**. For example,

the **matrix** $\begin{pmatrix} 6 & 1 & 3 \\ 4 & 2 & 7 \end{pmatrix}$ has three columns.

combination   An unordered arrangement. A combination from a **set** of n objects is any selection of n or fewer objects from the set, regardless of order.

For example, the full list of combinations of three letters from the set $\{a, b, c, d\}$ is: abc, abd, acd, bcd.

The number of combinations of r objects that can be made from a set of n objects is usually denoted by:

$$^nC_r \text{ or } \binom{n}{r}$$

It can be shown that:

$$^nC_r = \frac{n!}{r!(n-r)!} \quad \text{(where } n! \text{ means \textbf{factorial} } n\text{)}$$

For example, $^{12}C_5 = \dfrac{12!}{5! \times 7!} = 792$.

Notice that $^{12}C_5 = {}^{12}C_7$ and in general $^nC_r = {}^nC_{n-r}$.

The values of ^nC_r for varying r are the entries in the appropriate **row** of **Pascal's triangle.**

combined probability   The **probability** of an **event** that is combined in that it consists of more than one constituent event.

For example, the probability of a candidate passing an examination at the first attempt is 0.9. The probability of a candidate who has failed at the first attempt passing at the second attempt is 0.4. Finding the probability that a candidate will fail twice is finding a combined probability.

Problems of this kind are usually best solved using a **tree diagram**.

common denominator   A whole number that is a shared multiple of the **denominators** of two or more **fractions**.

In preparation for the **addition** or **subtraction** of fractions with unlike denominators, the fractions are usually converted into equivalent fractions with the same, or common, denominator:

$$\frac{1}{3}+\frac{1}{2}=\frac{2}{6}+\frac{3}{6}$$ 6 is the common denominator.

See **lowest**.

common difference   The difference between successive terms in an **arithmetic progression**. For example:

8, 11, 14, 17, 20   common difference 3
19, 17, 15, 13, 11   common difference -2

common logarithm   A **logarithm** to the **base** of ten.

commutative   Any **operation** $*$ which has the property $a*b = b*a$ for all members a and b of a given set is commutative.

Multiplication of real numbers is commutative. For example:

$$7\times3=3\times7.$$

Subtraction is not commutative. For example:

$$7-3\neq3-7.$$

complement The **set** of all the **elements** that are not in a particular set. Thus, the complement of a set S is the set consisting of the elements within the **universal** set that are not in S.

The complement of S is usually denoted by S'. Using the diagram:

> if ξ = {letters of the alphabet}
> and   C = {consonants}
> then C' = {vowels}.

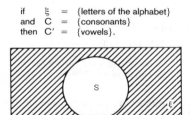

complement The shaded portion represents S'.

complementary angles Two **angles** whose **sum** is 90°.

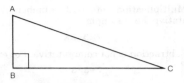

complementary angles In the right-angled triangle ABC $\angle$ BAC is the complement of $\angle$ BCA.

completing the square A method of solving a **quadratic equation** by reducing it to the form:

$$(x+h)^2=k$$

For example:

$$2x^2+12x-3=0 \Rightarrow x^2+6x-1.5=0$$
$$\Rightarrow x^2+6x=1.5 \Rightarrow (x+3)^2-9=1.5$$
$$\Rightarrow (x+3)^2=10.5 \Rightarrow x+3=\pm3.24$$
$$\Rightarrow x=0.24 \text{ or } x=-6.24$$

complex numbers An extension of the **real number** system designed originally to overcome difficulties involving the solution of problems like $x^2+3=0$, which have no solution in the **real numbers**.

Traditionally complex numbers are written in the form $a+bi$, where a and b are real numbers, and $i^2=-1$: a and b are called respectively the real and imaginary parts. $x^2+3=0$ has solutions $x=\sqrt{3}i$ and $x=-\sqrt{3}i$.

To add:

$$(a+bi)+(c+di)=(a+c)+(b+d)i.$$

To multiply:

$$(a+bi)\times(c+di)=(ac-bd)+(ad+bc)i.$$

Complex numbers have important implications in many branches of pure and applied mathematics. See **imaginary numbers**.

component   One of a **set** of two or more, often mutually **perpendicular**, **vectors** that together are equivalent in effect to the given vector.

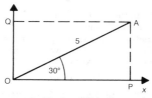

component   Vector **OP** in the x-axis and **OQ** in the y-axis.

In the illustration, if vector **OA** is of **magnitude** 5 and at 30° to the x-**axis** then:

Component in the direction of x-axis is vector **OP**.

Magnitude=5 cos 30°=4.33.

Component in the direction of y-axis is vector **OQ**.

Magnitude=5 sin 30°=2.5.

composite function   The combination of two or more **functions** to obtain a new one. The function $f:x \rightarrow (3x+1)^2$, can be thought of as the composition of two simpler functions:

$$g:x \rightarrow (3x+1) \text{ and } h:x \rightarrow x^2$$

$$x \xrightarrow{g} (3x+1) \xrightarrow{h} (3x+1)^2$$

$$f=hg$$

Notice the notation: $f=hg$ means f is equivalent to the function g followed by h.

Any function that can be split into two or more simpler functions in this way is called a composite function. For example:

if $f:x \rightarrow \sin^2 4x$,

then $f=pqr$, where $r:x \rightarrow 4x$, $q:x \rightarrow \sin x$
and $p:x \rightarrow x^2$.

compound interest   A method of calculating **interest** on money where the interest earned during a period is calculated on the basis of the original **sum** together with any interest earned in previous **periods**.

	1st year	2nd year	3rd year
Sum on which interest calculated	£100	£110	£121
Interest (@10%)	£ 10	£ 11	£ 12.10
Value of investment	£110	£121	£133.10

compound interest The appreciation of £100 invested at 10% per year for 3 years.

In general, if £P is invested at r per cent compound interest, then after n years it would be worth:

$$£P \times \left(\frac{100+r}{100} \right)^n$$

computer An electronic device designed to process large amounts of coded information rapidly.
   Computer systems commonly comprise three main elements:
(a) input devices (e.g. keyboard, floppy disk);
(b) central processor;
(c) output devices (e.g. monitor, printer).
   Rapid developments in microelectronic technology have resulted in drastic reductions in the size and cost of computers, whilst increasing their processing capabilities.

concave Describes a **curve** or surface that is hollow towards a given **point** of reference.

concave The dish of a radio telescope
presents a concave surface towards the
sky.

concentric circles Circles that have the same
centre but different **radii**.

concentric circles

concurrent Having a **point** in common. A

number of lines are concurrent if they all meet in a single point.

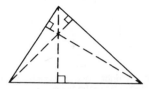

concurrent The three **altitudes** of a triangle are concurrent.

cone A **solid** bounded by a **plane** (usually circular) **base**, and tapering to a fixed **point** called the **vertex**. A *right cone* is one in which the axis, or line joining the centre of **symmetry** of the base to the vertex, is **perpendicular** to the base.

For a cone with a circular base, the volume of cone is:

$$V = \tfrac{1}{3}\pi r^2 h$$

where r=base radius, and h=height.

The curved surface area of a right cone is:

$$A = \pi r l$$

where l is the slant height.

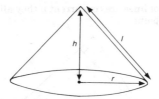

cone A right cone in which the axis is perpendicular to the base.

See **conic**.

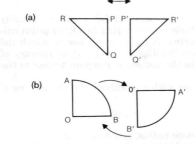

congruent (a) Reflection and (b) rotation map objects onto congruent images.

congruent Describes shapes, **plane** or **solid**,

that have the same shape and size.

Certain simple **transformations** such as **rotations**, **reflections** and **translations** map objects onto congruent images. See **mapping**.

conic A **curve** obtained by taking a **plane section** through a **cone**. By varying the **angle** of cut, four main types of conic can be obtained:

(a) **Circle**, cut **parallel** to the base.

(b) **Ellipse**.

(c) **Parabola**, cut parallel to the **slant**.

(d) **Hyperbola**.

Alternatively a conic can be thought of as the **locus** of a **point** that moves so that the **ratio** of its distance from a fixed point (the **focus**) to its distance from a fixed line (the **directrix**) is **constant**. This ratio is called the **eccentricity** of the conic.

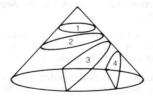

conic Four main types: (1) circle, (2) ellipse, (3) parabola and (4) hyperbola.

conjugate angle Two **angles** whose **sum** is 360°.

conjugate complex numbers Two **complex numbers** that differ only in the **sign** of their **imaginary** parts.

The conjugate of the complex number $a+bi$ is $a-bi$.

The **product** of a complex number and its conjugate is equal to the **square** of its **modulus**:

$$(a+bi)\times(a-bi)=a^2+b^2$$

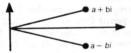

conjugate complex numbers A complex number and its conjugate related by a **reflection** in the real **axis** of the **Argand diagram**.

constant A fixed quantity in an expression. In the expression $y=3x+2$, 3 and 2 are constants whilst x and y are **variables**.

construction Additional **points** or lines to supplement a geometrical figure in order to prove some property of the figure.

continuity State of being continuous. Quantities such as length, weight, temperature, speed, time, etc., are *continuous* **variables**. For example, when measuring the height of pupils in a class the results are not **discrete**. In growing from 160 cm to 161 cm we assume every possible **value** between 160 and 161 has been attained.

In contrast, quantities like the number of pupils in a class can be measured by a counting process which varies in **integer** steps.

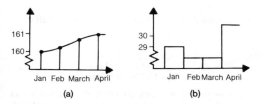

(a)   (b)

continuity A person's height is a continuous variable (a); the numbers of pupils in a class do not vary continuously (b).

converge To meet or join at a particular place. An **infinite sequence** that has a **limit** is said to converge or tend to the limit. This occurs when the difference between each term and the one following it becomes smaller throughout the sequence: that is the difference between the *n*th term and the $(n+1)$th term decreases as n increases.

When using a calculator or **computer** to check whether a sequence given **iteratively** converges it is usually sufficient to obtain a fairly small number of **elements** of the sequence, which may then be inspected for convergence. For instance, to determine whether the sequence:

$$x_{n+1} = \frac{3/x_n + x_n}{2} \quad \text{with } x_1 = 1$$

converges it is sufficient to obtain the results:

$$x_2 = 2$$
$$x_3 = 1.75$$
$$x_4 = 1.7321428 \text{ (7 dp)}$$
$$x_5 = 1.7320508 \text{ (7 dp)}$$
$$x_6 = 1.7320508 \text{ (7 dp)}$$

This sequence reaches its limit very quickly, as indicated by the results x_5 and x_6 being identical. See **diverge**.

converse A logical statement taken in reverse order. For example, the converse of the statement 'if a number ends in **zero** it is divisible by ten' is 'if a number is divisible by ten, then it ends in zero'.

Many important statements in mathematics are in the form 'if x then y'. The converse of such a statement is 'if y then x'.

The converse of a true statement need not

itself be true. For example 'if p and q are **even**, then $p+q$ is even' is a true statement. The converse 'if $p+q$ is even, then p and q are even' is false.

conversion The process involved in changing from one system (often of **units**) to another.

For example, to express a temperature, given in the **Fahrenheit** scale, in **centigrade**, use the conversion formula:

$$C=\frac{5}{9}(F-32)$$

convex Curved outwards. A curve or surface that bulges towards a given **point** of reference is called convex. The dome of St. Paul's Cathedral presents a convex surface to the sky.

A convex figure is one in which any line joining two points on the figure is contained within the figure and does not extend outside it.

convex A convex dome.

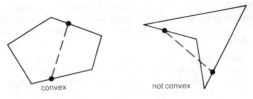

convex   not convex

convex Figures that are (a) convex and (b) not convex.

coordinates Numbers that define the position of points on a **plane** or in space. **Cartesian** and **polar** are two commonly used coordinate systems.

coplanar Lying in the same **plane**. Sets of **points** or lines lying in the same plane are called coplanar.

correlation A statistical term used to describe a relationship between two varying quantities, when changes in one **variable** are linked to changes in the other. A **positive** correlation occurs when increases and decreases in the two variables happen together. A **negative** correlation occurs when increases in one variable are associated with decreases in the other.

Doctors, for example, have suggested a positive

correlation between the habit of cigarette smoking and the incidence of heart disease.

A correlation **coefficient** is a number between −1 and 1, that gives a measure of correlation between the **values** of two **variables**.

correspondence A relationship involving the matching of pairs of **elements** from two **sets**.

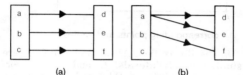

(a)   (b)

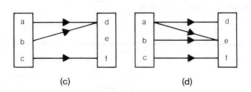

(c)   (d)

correspondence Four common types.

corresponding points, lines and angles When two figures are related by a simple **transformation**, features of the first figure are said to correspond to their images on the transformed figure.

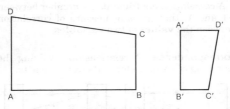

corresponding points and angles

In the illustration, ABCD and A′B′C′D′ are **similar quadrilaterals**, C and C′ are corresponding **points**, ∠ DAB and ∠ D′A′B′ are corresponding **angles**.

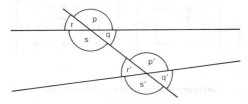

corresponding lines

When a **transversal** cuts a pair of lines, the pairs of angles, p and p′, q and q′, r and r′, s and s' are corresponding. If the pair of lines cut by the transversal are **parallel**, the corresponding angles are equal.

cosecant   A trigonometric function. In a **right-angled** triangle, the cosecant of an **angle** in the **ratio**, hypotenuse/opposite side:

$$\operatorname{cosec}\theta = \frac{1}{\sin\theta}$$

In △ ABC, the value of θ can be determined by examining a table of values for the cosecant function: when cosec θ = 1.25, θ = 53.1°.

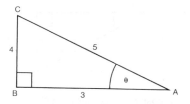

cosecant   In △ABC,

$$\operatorname{cosec}\theta = \frac{AC}{BC} = \frac{5}{4} = 1.25$$

cosine   A trigonometric function. In a **right-angled** triangle the cosine of an **angle** is the **ratio**, adjacent side/**hypotenuse**.

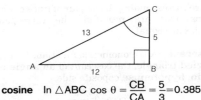

cosine   In $\triangle ABC$ $\cos \theta = \dfrac{CB}{CA} = \dfrac{5}{3} = 0.385$

In the $\triangle ABC$ the value of θ can now be determined by examining a table of values for the cosine function: when $\cos \theta = 0.385$, $\theta = 67.4°$.

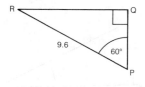

cosine   In $\triangle PQR$ $\cos 60° = \dfrac{PQ}{PR}$

From the table of values for the cosine function, $\cos 60° = 0.5$. Hence, in $\triangle PQR$:

$$0.5 = \frac{PQ}{9.6} \text{ and } PQ = 4.8$$

The cosine of an angle θ may also be defined as

the x coordinate of the point obtained when the point (1, 0) is rotated anticlockwise, about the origin, through an angle of θ.

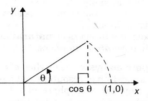

cosine   Defining the cosine of an angle.

cotangent   A trigonometric **function**. In a **right-angled** triangle the cotangent of an **angle** is the **ratio**, adjacent side/opposite side.

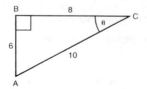

cotangent   In △ABC cot θ = $\dfrac{BC}{BA} = \dfrac{8}{6} = 1.33$

In the △ABC, the value of θ can now be

determined by examining a table of values for the cotangent function: when cot θ=1.33, θ=36.9°.

count   To match objects in a **set** in a one-to-one fashion with the names of the positive **integers** — 'one, two, three . . .'

The **positive** integers or **natural** numbers are often called *counting numbers*.

critical path analysis   A method of determining the best allocation of resources and most effective starting time for each phase of a large project. This is achieved by considering related activities with constraints on the order in which they are carried out. It is necessary because of restrictions on the availability of resources or because some tasks cannot be carried out until others are completed. The objective of the analysis is to ensure the project is completed as quickly as possible and with minimum expense.

cross-multiplying   A process of simplifying equations involving fractional terms. For example:

$$\text{If } \frac{4x}{3}=6, \text{ then } x=\frac{6\times3}{4} \quad \text{so } x=4\tfrac{1}{2}$$

cross-section   The **plane** shape that results from cutting through a **solid**, often at **right angles** to an **axis of symmetry** of the solid.

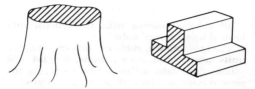

cross-section The cross-section of a tree trunk (left) and a girder (right).

cube 1. A **regular solid** with six **square faces**.
2. To raise a number to the **power** of three. 'Four cubed' is written

$$4^3 = 4 \times 4 \times 4 = 64.$$

See **cubic**, **cuboid**.

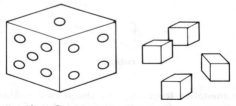

cube Common examples of cubes.

cubic 1. A cubic function is a **polynomial** of the third **degree**:

$$ax^3 + bx^2 + cx + d$$

All cubic equations with real **coefficients** have at least one real **solution**.

2. The cubic **metre**, cubic **centimetre**, etc., are **units** used in measuring **volume**. A **solid** with volume three cubic metres, written 3 m³, has the same volume as three **cubes** with 1 m **edges**.

See **cuboid**.

cuboid A **solid** with six **rectangular** faces, the opposite **faces** being equal in size. See **cube**, **cubic**.

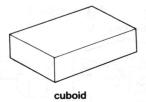

cuboid

cumulative frequency In **statistics** a method of grouping **frequencies** of the **value** of some **variable** by adding the frequencies not greater than certain values of the variable. A cumulative frequency graph is called an **ogive**.

Mark	Frequency
0	4
1	5
2	8
3	12
4	10
5	3

(a)

Mark	Cumulative frequency
≤0	4
≤1	9
≤2	17
≤3	29
≤4	39
≤5	42

(b)

cumulative frequency (a) Frequency table for pupils' test marks with (b) a cumulative frequency table for the same marks.

curve A line, either straight or continuously bending.

cusp A **point** on a **curve** where the curve turns back on itself, and where the two branches of the

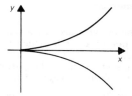

cusp A curve with a simple cusp at the origin; the x-axis is the common tangent.

curve share a common **tangent**, the branches being on opposite sides of the tangent. For example, $y^2=x^3$.

cyclic permutation A **permutation** of an ordered **set** achieved by **mapping** each object to its successor, the last object being mapped to the first. For example:

$$(1, 2, 3) \rightarrow (2, 3, 1) \rightarrow (3, 1, 2)$$

The first permutation is denoted by:

$$\begin{pmatrix} 1 & 2 & 3 \\ 2 & 3 & 1 \end{pmatrix}$$

cyclic quadrilateral A four-sided plane shape having **vertices** lying on the **perimeter** of a **circle**.

The opposite **angles** of a cyclic quadrilateral are **supplementary**.

cycloid The **curve** traced by a **point** P on the **circumference** of a **circle** which rolls along a straight line.

For a circle of radius a the parametric (see **parameter**) equation of a cycloid is:

$$x=a(\theta-\sin\theta) \quad y=a(1-\cos\theta)$$

See **epicycloid**.

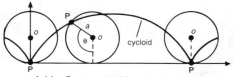

cycloid   Curve traced by point P.

cylinder   A **solid** with one **axis** of **symmetry** about which it has a uniform circular **cross-section**.

The volume of a cylinder is given by:

$V = \pi r^2 h$ (The **area** of a circular end multiplied by the height.)

The curved surface area of a cylinder is given by:

$A = 2\pi rh$ (The **circumference** of a circular end multiplied by the height.)

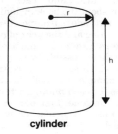

cylinder

D

data Information, often in numerical form, which has been collected for **statistical** purposes.

database A large collection of information stored as a number of records or files. Each record will normally consist of entries under various headings or fields. An example is a library catalogue where each book in the library has a separate record card, and each card contains information about the book including, perhaps as fields, the author's surname, first names, the publisher's name and the year of publication.

Computers can make particularly effective use of a database possible by very rapidly finding all records which satisfy a certain list of conditions. For example, a school's database consists of records, one for each pupil. Each record includes the pupil's surname, first names, date of birth, and the name of the teacher who teaches the pupil for each subject. The computer holding this information could be asked to find the names of all pupils aged 14 or over who are taught mathematics by a particular teacher.

decagon A **polygon** with ten sides.

decay rate The amount a particular quantity decreases in **unit** time. It is effectively the same as **negative growth rate**.

decimal A general term applied to the system of representing fractional and **whole numbers** in the **base** of ten.

A decimal **fraction** is one with **denominator** being a power of ten, for example, $\frac{3}{10}, \frac{7}{100}, \frac{38}{1000}$, etc.

The base 10 place **value** system for whole numbers is extended in the decimal system to include the representation of decimal fractions. The decimal point is used to identify place values:

$$13\frac{17}{100} = 13.17$$

tens tenths

units hundredths

All **rational numbers** can be represented in decimal form by a **terminating** or **recurring** sequence of decimal **digits**.

Irrational numbers can be represented approximately in decimal form by rounding off to a prescribed number of decimal places. For example:

$\sqrt{2} = 1.41$ to 2 decimal places
$\sqrt{2} = 1.4142$ to 4 decimal places
$\pi = 3.142$ to 3 decimal places.

decision tree A diagram which enables a decision to be reached by considering a sequence of questions each of which must be answered with either 'yes' or 'no'. See diagram opposite.

For example, to decide that a number is **prime** it is sufficient to check that no prime smaller than the square root of the number is a **factor** of the number. See **flow chart**.

degree 1. A **unit** for measuring **angles**, in which one complete **rotation** is divided into 360 degrees (written 360°).
2. The degree of a **polynomial** expression or **equation** of one **variable** is the highest **power** to which the unknown is raised. For example:

$7x^5 - 5x^3 + 19x^2 - 3$ is of degree 5 in x.

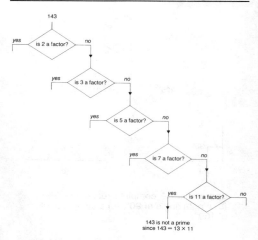

decision tree Procedure to check whether 143 is prime.

An expression in several variables, for example:

$$8x^3y^2z^4$$

is said to be of degree 9 (the sum of the powers of x, y and z), but of degree 2 in y.

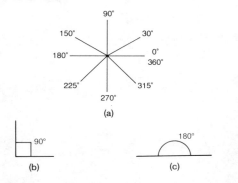

degree (a) A complete rotation of 360°;
(b) a right angle of 90°; (c) a straight angle
of 180°.

denominator The 'bottom number' of a **fraction**. For example, the denominator of $\frac{3}{4}$ is 4. See
common denominator, **lowest**.

depression The **angle** of depression of an
object from an observer viewing the object from
above, is the angle between the line joining the
object and the observer, and the horizontal
plane. See **elevation**.

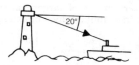

depression The angle of depression from the lighthouse to the ship is 20°.

derivative If $y=f(x)$ is a **function** of the **variable** x, the derivative of $y=f(x)$ at some $x=a$, is the rate of change of the function with respect to x at a. It is found by **differentiation**.

The derivative is often interpreted as the slope of the **graph** of $y=f(x)$ at $x=a$.

The derivative of $y=f(x)$ is usually written dy/dx or $f'(x)$. For $f(x)=x^3$, $f'(x)=3x^2$, thus the rate of change of x^3 at $x=4$ is $3\times4^2=48$.

determinant The determinant of a square **matrix M** is a number associated with that matrix.

For the 2×2 matrix $M=\begin{pmatrix} a & b \\ c & d \end{pmatrix}$ it is written

$\begin{vmatrix} a & b \\ c & d \end{vmatrix}$ or Det M, and is equal to $ad-bc$.

The determinants of larger matrices can be evaluated in a similar way. For example.

$$\begin{vmatrix} a & b & c \\ d & e & f \\ g & h & i \end{vmatrix} = a \begin{vmatrix} e & f \\ h & i \end{vmatrix} -b \begin{vmatrix} d & f \\ g & i \end{vmatrix} +c \begin{vmatrix} d & e \\ g & h \end{vmatrix}$$

The theory of determinants is particularly useful in the application of matrices to solving sets of **linear** equations, and linear **transformations**.

diagonal   A line joining any two non-adjacent **vertices** of a **polygon**.

An n sided polygon has $n-3$ diagonals from any vertex, and $\frac{1}{2}n(n-3)$ distinct diagonals in total.

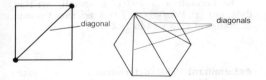

diagonal   Diagonal of (a) a **square** and (b) a **hexagon**.

diameter   A **chord** that passes through the

centre of a circle. The length of the diameter of a circle is twice the length of the **radius** of the circle.

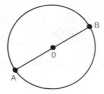

diameter If O is the centre of the circle then AB is the diameter.

difference The result obtained by subtracting one number from another, i.e. $a-b$.

The difference of two squares a^2-b^2 is the common name given to the factorization:

$$a^2-b^2=(a-b)(a+b)$$

The symmetric difference of two **sets** A and B is $(A\cap B')\cup(B\cap A')$ or $(A\cup B)\cap(A\cap B)'$ and is represented by the shaded area on the **Venn diagram** overleaf.

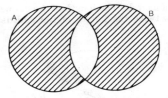

difference The symmetric difference of two sets.

differentiation The process of finding the **derivative** $f'(x)$ of a **function** f.

Essentially the process involves the evaluation of:

$$\text{Limit } \frac{f(x+h)-f(x)}{h} \text{ as } h \text{ tends to } 0.$$

This may be viewed as finding the **gradient** of the **tangent** to the **graph** of f at the point x, by taking the limit of the gradients of a sequence of **chords** over increasingly small intervals.

For many families of functions, derivatives are simply deducible from the form of the original function. For example:

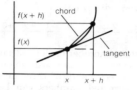

For $f(x) = Ax^n$, $f'(x) = nAx^{n-1}$
So, if $f(x) = 3x^3 + 2x^2 + 1$
then $f'(x) = 9x^2 + 4x$.

differentiation Finding the derivative $f'(x)$ of function f.

digit A single **numeral** (0, 1, 2, ... 9) when used as part of the representation of a number in a place-**value** system of writing numbers is called a digit. For example, 35 780 has five digits, the first being 3, and the last 0.

digital In numerical form. A digital **computer** is one that represents numbers by a digital code (usually of ones and **zeros**), rather than by a continuously measurable quantity as do *analogue* devices.

directed numbers Signed numbers (**positive** and **negative**) in which positive numbers are represented along a line relative to an origin 0, and negative numbers are similarly represented in the reverse direction:

$$\ldots \begin{array}{cccccccc} -3 & -2 & -1 & 0 & +1 & +2 & +3 & +4 \end{array} \ldots$$

In the directed number model, -3 can be thought of in two ways either as a position on the number line, or as a vector shift of 'three units left' along the line. Directed numbers can be used as a model for operations with **integers**. For example, $+1+-3=-2$ might be viewed as 'one step right then three steps left gives two steps left'.

directrix A line that defines the shape of a **curve** of the **conic** family. For any point on a given curve the **ratio** of its distance from a fixed **point** (**focus**) to the directrix line is **constant**.

discrete Quantities that can be measured by a counting process are called discrete. Discrete **variables** change in a discontinuous way. See **continuity**.

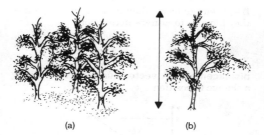

(a)   (b)

discrete (a) The number of trees in a plantation is a discrete variable over time; (b) the height of a particular tree is a **continuous** variable.

discriminant For a **quadratic equation** such as $ax^2 + bx + c = 0$ the quantity $b^2 - 4ac$ is called the discriminant. If:

$b^2 - 4ac > 0$ the equation has two real **roots**
$b^2 - 4ac = 0$ the equation has coincidental real roots
$b^2 - 4ac < 0$ the equation has **complex** roots.

disjoint Two **sets** are disjoint if their **intersection** is the **empty set**. If A = {**odd** numbers} and

B={**even** numbers} then A and B are disjoint — they have no **members** in common.

displacement A **vector** quantity representing a change of position.

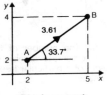

Displacement

As the **graph** shows the change of position from A(2,2) to B(5,4) can be represented by the vector $\binom{3}{2}$ and is of **magnitude** 3.61 at 33.7° to the x-axis.

distribution The set of **values** and associated **frequencies** of a **statistical variable**. There are various types of distribution, such as: **binomial**, **normal** and Poisson.

Number of sixes obtained in 10 throws of a die	Relative frequency
0	0.16
1	0.32
2	0.29
3	0.16
4	0.05
5	0.01
more than 6	0.01

(a)

distribution In throwing a die the number of sixes has a binomial distribution. (a) shown as a table, (b) shown as a **graph**.

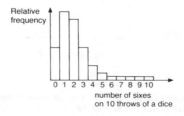

(b)

As the diagram below shows, statistical **data** such as the height of the girls in a class often has a **normal** distribution, whereas the number of telephone calls per minute passing through an exchange has a *Poisson* distribution:

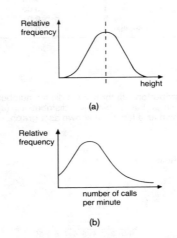

distribution (a) A normal distribution symmetric about the mean; (b) a Poisson distribution **skew**-symmetric.

distributive law When two **binary** operations $\oplus$ and $*$ are defined on a **set** S, and for any three elements x, y, z in S

$$x*(y\oplus z)=(x*y)\oplus(x*z)$$

$*$ is said to distribute over $\oplus$.

In the arithmetic of real numbers, multiplication distributes over addition. For example:

$$7\times(4+3)=(7\times4)+(7\times3)$$

Addition does not, however, distribute over multiplication. For example:

$$5+(3\times6)\neq(5+3)\times(5+6)$$

diverge An **infinite sequence** that has no **limit** is said to diverge. This occurs when the difference between each term and the one following either remains constant or increases throughout the sequence. (See **converge**.)

A divergent sequence may tend to infinity; for instance: 1, 10, 100, 1000, . . .

Alternatively it may oscillate, for instance: 2, -2, 2, -2, . . .

dividend A number that is to be divided by another. For example, in $2468\div3$, 2468 is the dividend. See **division**, **divisor**.

division The **inverse** operation to **multiplication**. One of the basic operations of **arithmetic**. The **quotient** of two numbers is determined by this operation which is associated with the process of dividing one quantity into a number of equal parts. See **dividend**, **divisor**.

divisor In a **division** problem the quantity by which the **dividend** is to be divided. The **factors** of a **whole number** are sometimes called divisors. For example, in 2468÷3, 3 is the divisor.

dodecagon A **polygon** having 12 sides, and hence 12 interior angles.

dodecagon

dodecahedron A **polyhedron** having 12 **faces**. A **regular** dodecahedron has for its faces 12 **regular pentagons**.

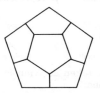

dodecahedron   A regular dodecahedron.

domain The **set** which acts as inputs to a **function** is called the domain of the function. For example, for the function of $y = \sqrt{x}$, the domain is normally taken to be the set of possible **values** of the **variable** x, i.e., non-negative **real numbers**.

dual In **plane** and **solid geometry** figures that can be obtained from one another by interchanging particular **elements** (for example, **points** ↔ lines) are called dual figures.

Theorems proved about one figure provide analogous theorems about the dual.

duodecimal The number system employing
the **base** of 12.

Two extra **numerals** more than the familiar 0,
1, 2, ... 9 are needed to represent 10 and 11 in
this system.

```
        Base 10              Duodecimal
          41          ↔         35
        /    \                /    \
     tens   units        twelves  units
```

If T and E are taken as symbols for 10 and 11
respectively then:

```
        Base 10              Duodecimal
          22          ↔         1T
          35          ↔         2E
          41          ↔         35
        /    \                /    \
     tens   units        twelves  units
```

E

e The **symbol** given to the **irrational** number which is the base of the natural **logarithm function**. A fundamental **constant** of both mathematics and the physical world.

> e=limit of $(1+1/n)^n$ as n increases.
> To ten places of decimals $e\approx2.7182818284$.

A unique property of the **function** $y=e$ is that the gradient of its **graph** for any **value** of x is identical to the value of the function for that value, i.e.

$$\frac{dy}{dx}=y$$

Thus the size of any quantity that grows or decays at a rate **proportional** to its size at a given time is related to the number e.

eccentricity For a **conic**, the ratio e of the distance of any **point** on the **curve** from the **focus**

to the distance of the same point from the **directrix**.

(a)

(b)

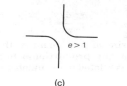

$e > 1$

(c)

eccentricity (a) For $e < 1$ the conic is an

ellipse; (b) for $e=1$ it is a **parabola**; (c) for $e>1$ it is a **hyperbola**.

edge A line forming the **intersection** of two **faces** of a **polyhedron** or other **solid**.

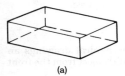

(a)   (b)

edge A **cuboid** (a) has 12 edges; a tetrahedron (b) has 6 edges.

element The individual members of a **set**. For example:

Red is an element of the set of primary colours.
7 is an element of the set of odd numbers.

The symbol ε means 'is an element of' in set notation, hence 7 ε {odd numbers}.

elevation 1. The **angle** of elevation of an object from an observer viewing the object from below, is the angle between the line joining the object and the observer, and the horizontal **plane**. See **depression**.

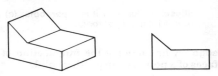

elevation An object and its side elevation.

2. In geometrical and mechanical drawings an elevation is a view of a **solid** object from the front or side of the object.

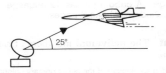

25°

elevation The angle of elevation of the plane from the radar station is 25°.

ellipse 1. An oval shape obtained by stretching or squashing a circle. More precisely, a plane **curve** belonging to the **conic** family. An ellipse has an **eccentricity** *e* of less than 1.

If a **cone** is cut by a plane to form a **cross-**

section which is a **closed** curve, the cross-section is an ellipse.

ellipse The cross-section of a cone forming an ellipse.

2. The **locus** of a **point** in a plane that moves such that the **sum** of the distances of the point from two fixed points, called the **foci**, is **constant**.

An ellipse has two **axes** of **symmetry** called the major and minor axes.

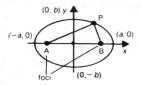

ellipse For any point, P, AP+BP=2a.

For an ellipse with major axis extending from $(-a, 0)$ to $(a, 0)$, and minor axis from $(0, -b)$ to $(0, b)$ the **Cartesian equation** is:

$$\frac{x^2}{a^2}+\frac{y^2}{b^2}=1$$

The foci are at $(c, 0)$ and $(-c, 0)$ where $c^2=a^2-b^2$ and the **directrices** are the lines

$$x=\frac{a^2}{c} \quad \text{and} \quad x=-\frac{a^2}{c}$$

A **circle** is a particular case of an ellipse where $e=0$. The area enclosed by an ellipse is equal to πab.

empty set Denoted by the symbol $\varnothing$, a set which contains no members. For example, the set of female kings of England $=\varnothing$.

enlargement A **transformation** of **plane** or **solid** shapes in which shapes are mapped onto **similar** shapes using a centre of enlargement.

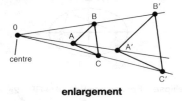

enlargement

$$\frac{A'B'}{AB} = \frac{B'C'}{BC} = \frac{A'C'}{AC} \text{ and } \frac{OB'}{OB} = \frac{OA'}{OA} = \frac{OC'}{OC}$$

The **scale factor** of an enlargement is the **ratio** of corresponding lengths on the object to those on the **image**. Alternatively the scale factor can be viewed as the ratio of distances between the centre and the object and corresponding distances between the centre and the image.

Enlargements centred at the origin can be represented in **matrix** form. For example, in a **plane**

$\begin{pmatrix} 3 & 0 \\ 0 & 3 \end{pmatrix}$ scale factor 3

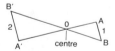

enlargement The matrix $\begin{pmatrix} -2 & 0 \\ 0 & -2 \end{pmatrix}$ scale factor 2 with a reflection in the centre.

envelope A **plane** curve which is **tangential**

to each member of a whole family of **curves**. In the diagram, the family of curves is of circles of radius a whose centres are all distance b from a fixed point 0.

> **envelope**   This family has two envelopes, themselves circles, one of radius $(b+a)$ and the other of radius $(b-a)$.

epicycloid   The path traced by a **point** P on the **circumference** of a **circle** which itself rolls on the circumference of another circle.

An epicycloid forms 'arches' around the fixed circle. If the **radii** of the rolling and fixed circles are a and b respectively, the curve has n 'arches' when $b=na$. See **cycloid**.

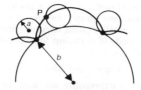

epicycloid The path traced by point P.

equal Quantities which are alike in certain respects are said to be equal in those respects. The symbol = is used to denote equality. The relationship of equality is an example of an **equivalence** relation.

equally likely events Events usually based on mathematical ideas like 'a fair die'. In practice, most dice are slightly unfair and tend to give slightly uneven results. A 'fair' die will tend to give exactly equal numbers of ones, twos, threes, fours, fives and sixes over a large number of throws. The events 'a one will come up' and 'a six will come up' are then said to be equally likely.

A football match may end as a home win, a draw or an away win. These are not equally likely events, as a glance at the results page will show. Home wins are most likely, and draws are least likely.

equation A statement of an equality relationship between two quantities or expressions. A large part of mathematics is concerned with the study of methods of **solution** for various types of equations. For example:

$$(x+1)^2 = x^2 + 2x + 1$$

is an equation expressing an identity relationship which holds for any numerical **value** of x.

$$3x + 1 = 10$$

is a simple **linear** equation which is conditional upon the value of x. The solution of this equation is $x=3$.

Other examples of common types of simple equations include **quadratic** and **simultaneous equations**.

equator A line of **latitude** which is a **great circle**. The equator is the **circle** produced on the earth's surface by taking a **section** through the **centre** of the earth, **perpendicular** to the earth's **axis**.

The equator is the reference line for positions of latitude on the earth's surface. For example, latitude 60° north means 60° north of the equator. See **small circle**.

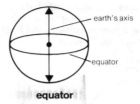

equator

equilateral   An equilateral **polygon** is one with
all its sides **equal** in length. An equilateral **triangle** also necessarily has equal interior angles.
This is however not generally true of all
equilateral polygons.

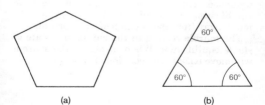

(a)   (b)

equilateral Equilateral pentagon (left)
and triangle (right).

equilibrium A state of balance. When the system of forces acting on a body are balanced and have a zero **resultant**, the body is said to be in a state of equilibrium.

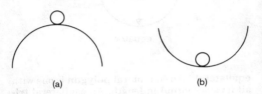

(a)   (b)

equilibrium A marble in (a) unstable equilibrium and (b) stable equilibrium.

In the diagram the marble, balanced on an upturned cup, is in a state of *unstable* equilibrium. When displaced the marble will move away from the position of equilibrium.

The marble resting in the cup is in a state of *stable* equilibrium. When displaced the marble will move back to the position of equilibrium.

equivalence relation A **relation** R between

the members of a **set** is said to be an equivalence relation if it has three properties:

(a) It is **reflexive**: cRc for all elements c in the set.

(b) It is **symmetric**: If aRb then bRa.

(c) It is **transitive**: If dRe and eRf then dRf.

For example, the relation 'is **similar** to' defined on a set of **plane** figures is an equivalence relation.

An equivalence relation splits up a set into distinct subsets called *equivalence classes*.

equivalent fractions Two **fractions** are called *equivalent* if they represent the same **rational** number, that is, they can be cancelled to produce a common fraction. For example $\frac{3}{4}$, $\frac{6}{8}$, $\frac{9}{12}$, $\frac{21}{28}$ and $\frac{315}{420}$ are all equivalent.

The fractions a/b and c/d are equivalent when $ad=bc$. See **cancellation**.

estimate To make a rough calculation, which may involve one or more **approximations**, so as to obtain an inexact but helpful preliminary **solution** to a problem.

For example, to estimate that $1369-284$ is about 1100 may be done by considering that 1369 is approximately 1400 (to the nearest 100) and that 284 is approximately 300. Then $1400-300=1100$.

To estimate that $352 \div 69$ is about 5 may be done by considering that 352 is approximately 350 (to the nearest 10) and 69 is approximately 70. Then $350 \div 70 = 5$.

Euclidean A term often used to describe the **geometry** of ordinary space of two or three dimensions.

(Named after the 3rd century BC Greek mathematician Euclid, who devised the principles of geometry.)

Euler's formula 1. The formula that relates the number of **vertices**, **faces** and **edges** in a **polyhedron**, i.e.

faces + vertices = edges + 2

2. The formula that relates the number of **nodes**, **regions** and **arcs** in a **network**, i.e.

nodes + regions = arcs + 2

(Named after the Swiss mathematician and physicist Leonhard Euler (1707–83), famous for his work in geometry, calculus and trigonometry.)

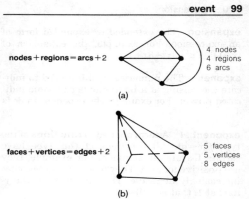

nodes + regions = arcs + 2

4 nodes
4 regions
6 arcs

(a)

faces + vertices = edges + 2

5 faces
5 vertices
8 edges

(b)

Euler's formula (a) For networks (b) for polyhedra.

even Numbers that have 2 as a **factor** are called even. Even numbers are the multiples of 2: 2, 4, 6, 8 . . .

event A clearly defined occurrence, such as 'this coin will come up heads'. Some events are impossible, others are certain. Most fall somewhere in between and are (less or more) likely. The measure of this likelihood is **probability**. See **equally likely events**.

expansion The extended or expanded form of an expression. For example, the expansion of $(x+1)^3$ is x^3+3x^2+3x+1.

exponent The numerical **symbol** used to indicate the raising of a **base** number to some indicated **power**. For example, the exponent in 3^4 is 4.

exponential A term applied to **functions** of the type $f(x)=a^x$. All functions of this type have certain features in common. For example, they are **positive** for all x. They tend to 0 as x becomes increasingly **negative**. Their **graphs** cut the y-axis at 1, that is, $f(0)=1$.

The function $f(x)=e^x$, in particular, is known as the exponential function. These functions are associated with situations of natural growth and decay. Their inverses are logarithmic functions.

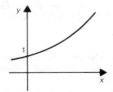

exponential The function cuts the y-axis at 1.

extrapolate   To use past figures to try to predict the future. In the graph we try to predict the population of Great Britain in 1911 from the figures over the period 1821–91.

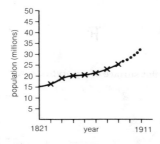

extrapolate   Predicting population figures.

F

face The flat surface of a **polyhedron**.

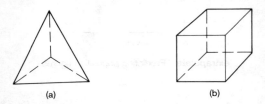

(a)   (b)

face A pyramid (a) has four faces while a
cube (b) has six faces.

factor All the numbers which divide exactly
into a number. For example:

factors of 12 are 1, 2, 3, 4, 6, 12
factors of 17 are 1, 17
factors of 32 are 1, 2, 4, 8, 16, 32

The factors of a **polynomial** are other polynomials that divide into it exactly. For example:

factors of x^3-8 are $x-2$ and x^2+2x+4
factors of x^3+1 are $x+1$ and x^2-x+1

factor theorem A **theorem** that states that if P(x) is a **polynomial** and P(a)=0, then $x-a$ is a **factor** of the polynomial. For example:

if $P(x)=x^3-6x^2+11x-6$
$P(1)=1-6+11-6=0 \Rightarrow x-1$ is a factor
$P(2)=8-24+22-6=0 \Rightarrow x-2$ is a factor
$P(3)=27-54+33-6=0 \Rightarrow x-3$ is a factor

so the factors of $x^3-6x^2+11x-6$ are $(x-1)$, $(x-2)$ and $(x-3)$ and

$$x^3-6x^2+11x-6 \equiv (x-1)(x-2)(x-3).$$

factorial Factorial n, written as $n!$ is defined as the product of all the **integers** 1, 2, 3, ... up to and including n. For example:

$5! = 5 \times 4 \times 3 \times 2 \times 1 = 120$
$4! = 4 \times 3 \times 2 \times 1 \quad = 24$
$3! = 3 \times 2 \times 1 \quad\quad = 6$
$2! = 2 \times 1 \quad\quad\quad = 2$
$1! = 1 \quad\quad\quad\quad = 1$

Using the above pattern, we may deduce a value of 1 for 0!=1.

factorize To express a number or **polynomial** as the **product** of some of its **factors**. For example:

$$12=6\times2=4\times3=2\times2\times3,$$
$$125=5^3=5\times25,$$
$$x^3-8=(x-2)(x^2+2x+4),$$
$$x^3-2x^2+x=x(x-1)^2.$$

Fahrenheit scale A scale of temperature measurement. On the Fahrenheit scale the boiling and freezing points of water are 212° and 32° respectively. The normal body temperature of a person is 98.4° Fahrenheit (98.4°F).

A temperatue in Fahrenheit may be changed to a temperature in **centigrade**, the other common scale of measurement, using the formula:

$$C=5/9 \ (F-32).$$

Fibonacci numbers A sequence of numbers in which each member of the sequence is the sum of the two preceding numbers: 1, 1, 2, 3, 5, 8, 13, 21, 34 etc.

(Named after the Florentine mathematician Leonardo Fibonacci (*c.* 1170–1250) who introduced the Arabic number system to Europe.)

flowchart A diagram showing the procedure to be followed in solving a problem. For example, an **iterative** way of finding $\sqrt[3]{10}$ is shown by the flowchart.

See **decision tree**.

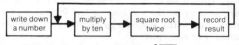

flowchart A way of finding $\sqrt[3]{10}$.

focus A **point** associated with a **curve** of the **conic** family. For any point on a conic curve, the **ratio** of the distance of the point from the focus to its distance from the **directrix** line is a **constant** (called the **eccentricity**).

foot A **unit** of length. 1 foot is equal to 12 **inches** or just over 30 **centimetres**. There are 3 feet in a **yard**.

formula A general rule or result stated in algebraic **equation** form. For examples:
(a) $P=2(l+w)$ is a formula for the **perimeter** of a **rectangle**, l and w stand for the length and width of the rectangle.

(b) $F = 9/5C + 32$ is a formula linking temperatures in degrees **Fahrenheit** (F) and **centigrade** (C).

(c) $A = \pi R^2$ is a formula for the area of a circle in terms of its **radius** R.

In the above examples the letters P, R and A are said to be the *subjects* of the formulae. We may change the subjects by rearranging the formula. For example:

$$P = 2(l + w)$$
$$\frac{P}{2} = l + w$$
$$\sqrt{\frac{P}{2}} - l = w$$

w is now the subject

$$A = \pi R^2$$
$$\frac{A}{\pi} = R^2$$
$$\sqrt{\frac{A}{\pi}} = R$$

R is now the subject

fraction   Part of a whole. The fraction 'three-quarters' is denoted by the **rational number** $\frac{3}{4}$. The bottom number is called the **denominator** and the top number is called the **numerator**.

The examples below use fractions.

Addition:

$$4\frac{5}{8} + 3\frac{7}{12} = 7\frac{15 + 14}{24} = 7\frac{29}{24} = 8\frac{5}{24}$$

Subtraction:

$$4\frac{3}{8} - 1\frac{7}{12} = 3\frac{9}{2}\frac{24 - 14}{24} = 2\frac{19}{24}$$

fraction The shaded part represents $\frac{5}{8}$ of the whole circle, which has been divided into eight equal parts.

Multiplication:

$$3\tfrac{3}{4} \times 1\tfrac{1}{9} = \frac{\overset{5}{15}}{4_2} \times \frac{\overset{5}{10}}{9_3} = \frac{25}{6} = 4\tfrac{1}{6}$$

Division:

$$3\tfrac{3}{4} \div 1\tfrac{1}{9} = \frac{15}{4} \div \frac{10}{9} = \frac{\overset{3}{15}}{4} \times \frac{9}{10_2} = \frac{27}{8} = 3\tfrac{3}{8}$$

A fraction may be changed into a **decimal** by dividing the numerator by the denominator. For example:

$$\tfrac{3}{8} = 0.375 \qquad \frac{0.375}{8)3.0^60^40}$$

frequency The number of times an **event** has occurred. For example, a die is thrown ten times and the results are 1, 2, 5, 6, 5, 4, 3, 1, 2, 1. The frequency of 1 is three since three of the throws were 1s. The results may be written in a frequency table:

Score		1	2	3	4	5	6
Number of throws (frequency)		3	2	1	1	2	1

The relative frequency of 5 is

$$\frac{\text{frequency of 5}}{\text{total number of throws}} = 0.2$$

frustum Any part of the **solid** contained between two **parallel planes** that cut the solid.

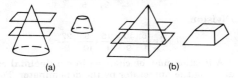

(a)   (b)

frustum The frustum of (a) a **cone** and (b) a **pyramid**.

function A relation between two **sets** called the **domain** and **range** in which each member of the domain is related to precisely one member of the range, called its **image**.

For a function f and **element** x in the domain of f, the image of x is denoted by $f(x)$. x and $y=f(x)$ are **variables**. For example:

$x \rightarrow 2x+3$ Describes a function from the set of **real numbers** to itself.

$x \rightarrow$ 'number of factors of x'

Describes a function from the set of positive **integers** to itself.

$x \rightarrow 1/x$ Describes a function from the set of non-**zero** real numbers (0 would have no image) to itself.

G

gallon A measure of **volume**. There are 8 **pints** in 1 gallon. 1 gallon is approximately 4.5 **litres**.

geometric mean In a **set** of n numbers the nth **root** of the **product** of all of the numbers. For example:

geometric mean of 2, 4, 8 is

$$\sqrt[3]{2\times4\times8}=\sqrt[3]{64}=4$$

geometric mean of 2, 3, 6, 12, 18, is

$$\sqrt[5]{2\times3\times6\times12\times18}=6$$

geometric progression A **sequence** of numbers, in which each term is a constant **multiple** of the preceding term. This multiple is called the *common* **ratio**. For example:

> 2, 6, 18, 54, 162   common ratio is 3
> 64, 32, 16, 8, 4, 2   common ratio is $\frac{1}{2}$
> $\frac{1}{4}$, -1, 4, -16, 64   common ratio is -4

The sum of the terms of the general geometric progression $a+ar^2+ar^3 \ldots +ar^{n-1}$ is given by the **formula**:

$$S_n = \frac{a(r^n - 1)}{r-1} \quad \text{or} \quad \frac{a(1-r^n)}{1-r}$$

(a is the first term of the series and r is the common ratio).

The sum of an infinite geometric progression can be calculated if the common ratio is between -1 and $+1$.

$$S_\infty = \frac{a}{1-r}$$

geometry A branch of mathematics which involves the study of properties of **points**, lines and **planes** and of **curves**, shapes and **solids**.

giga- A prefix with **symbol** G which stands for one thousand million (10^9). For example, 1 gigabyte is equal to one thousand million bytes.

glide reflection A geometrical **transformation** of the plane composed of a **reflection** and a **translation**. In the diagram, $\triangle A'B'C'$ is obtained from $\triangle ABC$ by a glide reflection in the x-axis. In **vector** notation:

$$\begin{pmatrix} x \\ y \end{pmatrix} \rightarrow \begin{pmatrix} x+2 \\ -y \end{pmatrix}$$

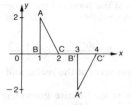

glide reflection △A′B′C′ is obtained from △ABC.

gradient A measure of steepness of a line or **curve**.

The gradient of a curve at a particular point is the gradient of the **tangent** drawn to the curve at that point. The curve $y=f(x)$ has gradient $f'(x)$.

gradient Lines sloping this way have a positive gradient. $\frac{4}{2}=2$.

gradient Lines sloping this way have a negative gradient. $\frac{-5}{2}=-2\frac{1}{2}$.

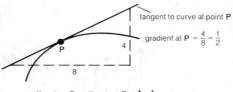

tangent to curve at point P

gradient at P $=\frac{4}{8}=\frac{1}{2}$

gradient Gradient at P$=\frac{4}{8}=\frac{1}{2}$.

gram A **unit** of mass. There are one thousand grams in a **kilogram** (1000 g=1 kg), and just over 28 grams in 1 **ounce**. 1 **cubic centimetre** of water has a mass of 1 gram.

graph A diagrammatic representation of the relationship between various quantities. Often, **coordinates** and **axes** are used. Statistical information is often displayed in a **bar chart**.

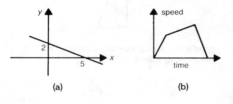

(a)

(b)

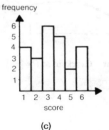

(c)

graph (a) Graph of the equation $2x+5y=10$; (b) a speed–time graph; (c) a bar chart.

great circle   A **circle** drawn on the earth's surface that has the same **radius** as that of the earth. For example, the **equator** is a great circle; all **longitude** lines are parts of great circles. See **small circle**.

great circle   The shortest distance along the surface between two points is along a great circle route.

group   A **set** of numbers, letters, etc., is said to form a group under a **binary operation** ∗ if the four following properties hold:
(a) The set is **closed** under the operation ∗.
(b) The operation ∗ is **associative**.
(c) An **identity** element exists and is unique.
(d) Each **element** of the set has a unique **inverse**.

For example, all 2 by 2 **matrices** with non-**zero determinant** form a group under matrix multiplication.

The numbers 1, 5, 7, 11 form a group under multiplication **modulo** 12.

growth rate The amount a particular quantity increases in **unit** time. For example, the population of the UK is not fixed, and tends to grow, having been about 30 million in 1900. The population changes constantly because people are constantly being born and others are dying. The increase in population for a particular period is found by taking the number of births during the period and subtracting the number of deaths during the period. The growth rate is then given by the increase divided by the time. For instance, if the increase in a population in one year were found to be 1.2 million then the growth rate would be given as 1.2 million per year.

This reasoning leads to a differential **equation**, if P, the population, is considered to be a function of time:

$$\frac{dP}{dt} = (B - C)P$$

where B is the birth rate and C is the death rate.

H

half-life The time taken for the activity of a radioactive isotope to decay to half of its original value; that is, for half of the atoms present to disintegrate.

hectare An **area** of 10 000 **square metres**. A **square** of side 100 m has an area of 1 hectare.

helix A **curve** which lies on the surface of a **cylinder** or **cone** and makes a constant **angle** with the **axis**. See diagram overleaf.

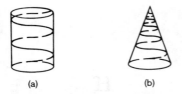

(a)   (b)

helix The curve on the surface of (a) a cylinder and (b) a cone.

hemisphere Half of a **sphere**. It is formed when a sphere is cut into two, the cut passing through the sphere's **centre**.

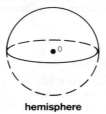

hemisphere

heptagon A seven-sided **polygon**. The sum of all the interior angles of a heptagon is 900°.

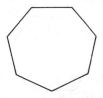

heptagon A **regular** heptagon (all sides equal).

Heron's formula A **formula** for the **area** of a **triangle** using the lengths of its sides.

$$A=\sqrt{s(s-a)(s-b)(s-c)}$$

where s is the semi-perimeter $\frac{1}{2}(a+b+c)$.

(Named after the 1st century Greek mathematician Heron of Alexandria.)

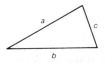

Heron's formula.

hexagon A **polygon** having six sides. A **regular** hexagon is one whose interior angles are each equal to 120°.

hexagon A regular hexagon.

highest common factor (HCF) Of two or more numbers, the largest number that divides exactly into each of them. For example:

HCF of 6, 8, 12 is 2
HCF of 10, 15, 20 is 5
HCF of 20 and 45 is 5
HCF of 42 and 56 is 14

histogram Similar to a **bar chart** except that the **frequency** of the bar is represented by its **area** rather than its height.

Length of phone call (minutes)	Number of calls	Calls/minute
0–1	10	$10 \div 1 = 10$
1–2	15	$15 \div 1 = 15$
2–5	15	$15 \div 3 = 5$
5–10	20	$20 \div 5 = 4$

(a)

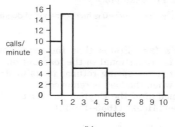

(b)

histogram The information (a) converted into a histogram (b).

homogeneous   polynomial A polynomial whose terms are all of the same **degree**.

$x^3+x^2y+3xy^2+7y^3$ is homogeneous of degree 3

or

$2x^2+3xy-y^2$ is homogeneous of degree 2.

Note: the degree of the term is obtained by adding together the powers of x and y, so x^3y^4 is of degree 7.

Similarly a homogeneous **equation** may be defined. For example:

$x^3+x^2y+3xy^2+7y^3=0$ is homogeneous of degree 3.

Hooke's law States that the extension of a spring is proportional to the load put on it (provided that the spring returns to its unstretched length when the load is removed).

(Named after the English physicist and chemist Robert Hooke (1635–1703).)

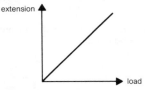

Hooke's law $E \propto L$ or $E = k \times L$ where k is a constant depending on the properties of the material of the spring.

horizontal   A line is said to be horizontal if it is **parallel** to the earth's skyline. A liquid which is at rest has a horizontal surface.

horizontal Examples of lines parallel to the earth's skyline.

hour   A **unit** of time. 1 hour=60 **minutes**, and there are 24 hours in a day.

hyperbola   A **curve** of the **conic** family. It is the **locus** of a **point** which moves so that the **ratio** of its distance from a fixed point (called the **focus**) to its distance from a fixed line (called the **directrix**) is a **constant** greater than 1. The ratio:

$$\frac{PF_1}{PQ}=e>1$$

The ratio e is called the **eccentricity** of the hyperbola.

A hyperbola has two foci (plural of focus). The two dotted lines in the diagram overleaf are called the **asymptotes** of the hyperbola. The **Cartesian equation** of the hyperbola drawn is

$$\frac{x^2}{a^2} - \frac{y^2}{b^2} = 1$$

It has a focus $(a, 0)$ and a directrix line $x = a^2/c$ where $c^2 = a^2 + b^2$. The asymptotes are the lines

$$\frac{y}{b} = \frac{x}{a} \text{ and } \frac{y}{b} = \frac{-x}{a}$$

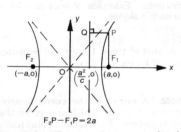

$F_2P - F_1P = 2a$

hyperbola The asymptotes of the hyperbola.

hypocycloid The **locus** of a point P fixed on the **circumference** of a **circle** which rolls on the inside of a given fixed circle. The points where P touches the circle are called **cusps**.

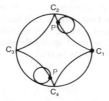

hypocycloid C_1, C_2, C_3, C_4 are cusps. The fixed circle has a diameter four times that of the rolling circle.

hypotenuse The longest side of a right-angled **triangle**. The hypotenuse is always opposite the right angle.

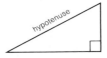

hypotenuse The side opposite the right angle.

hypothesis A statement or theory which is believed to be true, but which has not been proved. Statistical work, sometimes based on surveys, may be used in establishing whether a particular hypothesis is true.

I

icosahedron A **polyhedron** having 20 **faces**. A **regular** icosahedron is made up of 20 **equilateral** triangles. See diagram overleaf.

identity An identity of a **set** S with **binary operation** $*$ is a member i of the set which, when combined with any other member of the set, leaves it unchanged: $i*x=x*i=x$ for $x \, \varepsilon \, S$.

For example, for **addition**, 0 is the identity since $0+7=7$, $0+-2=-2$, $0+150=150$, etc.

For **multiplication**, 1 is the identity since $1\times7=7$, $1\times-2=-2$, $1\times19=19$, etc.

For sets under the operation of **union**, $\varnothing$ (the **empty set**), is the identity:

$$\varnothing\cup\{a, b, c, d\}=\{a, b, c, d\}. \; \varnothing\cup A=A \text{ (for any A)}.$$

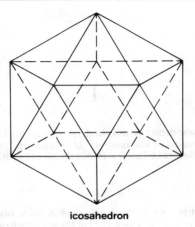

icosahedron

For 2×2 **matrices** the identity for multiplication is:

$$\begin{pmatrix} 1 & 0 \\ 0 & 1 \end{pmatrix} \text{ since } \begin{pmatrix} 1 & 0 \\ 0 & 1 \end{pmatrix} \begin{pmatrix} a & b \\ c & d \end{pmatrix} = \begin{pmatrix} a & b \\ c & d \end{pmatrix}$$

In the operation illustrated, b is the identity. Note: the 'b' row is the same as the top row of the

table, and the '*b*' column is the same as the left
column of the table.

identity   *b* is the identity here.

image   The result of applying a **function** to a
particular member of the **domain** is called the

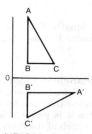

image   △A'B'C' is the image of △ABC
after a 90° **rotation**.

image of that **element**. For example, $x \rightarrow 2x+3$, $2 \rightarrow 7$. The number 7 is said to be the image of the number 2 which is called the *object*.

In **geometry** the term is often applied to the result of some **transformation**.

imaginary number The part of a **complex number** which in $a+bi$ is b. For example, the imaginary part of $3+4i$ is 4.

improper fraction A **fraction** whose **numerator** is greater than its **denominator**. For example, $\frac{17}{6}$, $\frac{8}{3}$, $\frac{13}{4}$, etc.

An improper fraction may be changed into a **mixed number**. For example, $\frac{17}{6}=2\frac{5}{6}$, $\frac{13}{4}=3\frac{1}{4}$.

inch A **unit** of length. There are 12 inches in 1 **foot** and 1 inch is just bigger than 2.5 **centimetres**.

incircle In a **triangle** in which the **angles** have been **bisected**, all these bisectors meet at the same point 0, called the *incentre* of the triangle. With 0 as **centre** it is possible to draw a circle that just touches each of the sides of the triangle. This circle is called the incircle, or **inscribed** circle of the triangle.

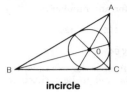

incircle

independent variable The equation $y=2x+5$ defines y as a function of x. x is said to be the independent **variable** of the **function** (and y is the *dependent variable*).

In the equation $S=2t^2+3t-5$, the independent variable is t.

index (plural, indices) In the number 2^3, 3 is the index or **power**. When numbers are multiplied or divided, the rules below apply:

(a) $x^a \times x^b = x^{a+b}$ (add indices when multiplying)

(b) $x^a \div x^b = x^{a-b}$ (subtract indices when dividing)

(c) $(x^a)^b = x^{ab}$

Using these rules, negative and fractional indices may be deduced:

$$x^{-a} = \frac{1}{x^a},\ x^{1/n} = \sqrt[n]{x},\ \text{and}\ x^{a/b} = \sqrt[b]{x^a}\ \text{or}\ (\sqrt[b]{x})^a$$

also $x^0 = 1$ for any number x

$$2^{-5} = \frac{1}{2^5} = \frac{1}{32}, \quad 125^{1/3} = \sqrt[3]{125} = 5,$$

$$8^{2/3} = \sqrt[3]{8^2} = \sqrt[3]{64} = 4.$$

When a number (for example, 16) is written as $16 = 2^4$, it is said to be written in index form. Similarly,

$$360 = 2 \times 2 \times 2 \times 3 \times 3 \times 5$$
$$= 2^3 \times 3^2 \times 5^1.$$

inequality Also known as *ordering*, a mathematical statement that one quantity is greater or less than another. For example:

$3 > 2$ (3 is greater than 2)
$1 < 4$ (1 is lesser than 4)
$x \geq 1$ (x is greater than or equal to 1)
$y \leq 2$ (y is less than or equal to 2)

Inequalities remain true if the same quantity is added to or subtracted from both sides, or if both sides are multiplied or divided by the same **positive** number. However, multiplying or dividing by a **negative** number changes the sense of the inequality:

$5 > 2$ adding 3 $8 > 5$
   subtract 4 $1 > -2$ are all still true
   $\times$ by 3 $15 > 6$
   $\div$ by 2 $2\frac{1}{2} > 1$

but

| × by −4 | −20 < −8 | direction of sign |
| ÷ by −2 | −2½ < −1 | must change to make the statement true. |

See **inequation**.

inequation A statement involving **inequalities**. For example:

$$x^2 + 2x + 2 > 0$$

holds for any numerical value of x.

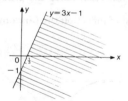

inequation $y < 3x − 1$ is satisfied by all points below the line $y = 3x − 1$.

infinite A quantity is said to be infinite if it is larger than any fixed limit. For example, the **set**, $\{1, 2, 3, 4, \ldots\}$ has an infinite number of members. But the set, {a, b, c, ... y, z} has a finite number of **members**, since they can be counted (26 members).

The symbol ∞ is used to represent an infinite **value**, or infinity.

inflection A point on a **curve** where the curve changes from being **concave** upwards to concave downwards (or vice versa). In the diagram, point P is said to be a **stationary point** of inflection since the **gradient** at P is zero. Q is a non-stationary inflection since the gradient is non-zero at Q.

At a point of inflection the **tangent** crosses the curve.

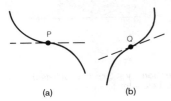

(a)   (b)

inflection (a) Gradient is zero at P; (b) non-zero gradient at Q.

inscribed circle A **circle** within a **polygon** in

which each side of the polygon is **tangent** to the circle. See **incircle**.

inscribed circle The circle in the triangle has its **centre** at the point of intersection of the angle bisectors.

integer A **whole number**. For example:

... −5, −4, −3, −2, −1, 0, +1, +2, +3, +4, +5 ...
negative integers   **positive** integers

integral The result of **integration**. The integral of the **function** $f(x)$ is defined as the limit as δx approaches zero, of the sum of the **areas** of the **rectangles** under the **curve**. The area under the curve between the values $x=a$ and $x=b$ is written as:

$$\int_a^b f(x)\ \mathrm{d}x$$

This is called a *definite integral* and a and b are known as the *lower* and *upper* limits of the integral.

Any function whose **derivative** is $f(x)$ is called an *indefinite integral* of $f(x)$. For example, $3x^2+7$, $3x^2-14$, $3x^2+K$ are all indefinite integrals of $6x$.

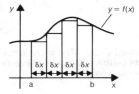

integral   The limit of the sum of the areas of the rectangles under the curve.

integration   The process of finding a definite or indefinite **integral**. Integration may be thought of as the inverse of **differentiation**.

Some important results are:

$$\int x^n dx = \frac{1}{n+1}x^{n+1}+K \quad (\text{for } n \neq -1)$$

$$\int \sin x\, dx = -\cos x \pm K$$

$$\int \cos x\, dx = \sin x + K$$

intercept   Of a straight line on an **axis**, the distance from the origin to the point where the line cuts the axis.

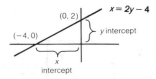

intercept   The x and y intercepts are -4 and 2 respectively.

interpolation   The process of finding the **value** of a **function** at a point using two known **values** on either side of the point.

For example, using linear interpolation the predicted value of $f(c)$ using the known values $f(a)$ and $f(b)$, is shown on the diagram.

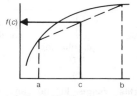

interpolation   Finding the value of $f(c)$.

interquartile range (IQR) Of a distribution of numbers, the difference between the upper and lower **quartiles** of the distribution.

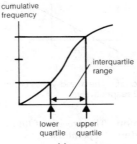

(a)

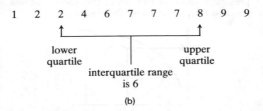

(b)

interquartile range In (a) and (b) IQR = (upper quartile) − (lower quartile).

intersection The set of points in common to two lines or curves.

The intersection of two sets is the set of elements that they have in common. The **symbol** ∩ means intersection.

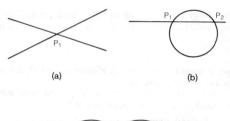

(a)

(b)

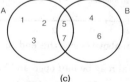

(c)

intersection (a) Two lines intersecting in a single point; (b) a line intersecting a circle in two points; (c) the intersection of two sets: A={1, 2, 3, 5, 7}, B={4, 5, 6, 7}⟹A∩B={5, 7}

interval If x is a **real number** and can take all **values** between -2 and $+3$ inclusive, we say that x lies in the interval -2 to $+3$. This may be represented in a diagram using a number line:

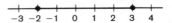

The diagram above shows the 'end points' are included. Algebraically this interval is written as:

$$\{x: -2 \leqslant x \leqslant 3\} \text{ or } [-2, 3]$$

and is sometimes called a **closed** interval, since both end points are included.

An interval where the end points are not included is called an *open* interval:

The open interval: $\{x: -1 < x < 5\}$ or $]-1, 5[$

The diagram above shows the 'end points' are not included.

An interval may be half open and half closed:

$$\{x: -2 \leqslant x < 4\} \text{ or } [-2, 4[$$

The diagram above shows -2 included in the interval and 4 excluded.

invariant A quantity or property is invariant under a **transformation** if it remains unchanged by the transformation. For example, the areas of **plane** figures remain invariant under certain transformations such as **rotations**, **reflections**, **shears**, **translations**.

inverse If a **set** with a **binary** operation $*$ defined on it has an **identity** element i, then any element x for which there exists another element x^{-1} with $x*x^{-1}=i$ is said to have an inverse element. For example:

(a) The inverse of $+7$ under addition is -7.

(b) The inverse of 8 under multiplication is $\frac{1}{8}$.

(c) The inverse of the **matrix** $\begin{pmatrix} 5 & 3 \\ 6 & 4 \end{pmatrix}$ under

multiplication is $\begin{pmatrix} 2 & -1\frac{1}{2} \\ -3 & 2\frac{1}{2} \end{pmatrix}$.

(d) The inverse of the **function**

$f(x)=\dfrac{x-1}{2}$ is $f^{-1}(x)=2x+1$.

The inverse of the plane **transformation** 'rotate through $+90°$ about $(0, 0)$ is 'rotate through $-90°$ about $(0, 0)$'.

inversely Two quantities vary inversely if one decreases while the other increases, their **product** remaining **constant**. They are said to be inversely **proportional** to each other.

involute The **curve** of a **circle** that is described by the end of a thread as it is unwound from a fixed spool (keeping the thread taut all the time).

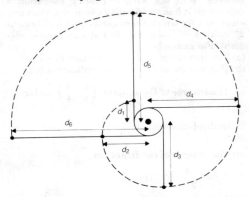

involute The distances d_1, d_2, d_3, . . . are equal to $\frac{1}{4}$, $\frac{1}{2}$, $\frac{3}{4}$, etc. of the **circumference** of the circle.

irrational number A number that is not **rational**, that is, it cannot be expressed as a **fraction**, or written in exact form as a finite **decimal**. For example, π, e, $\sqrt{2}$, $\sqrt{3}$, $1-\sqrt{7}$. All the irrational numbers together with all the rational numbers make the set of real numbers.

It can be shown that $\sqrt{2}$ is an irrational number as follows:

Suppose $\sqrt{2}$ is rational

i.e. $\quad \sqrt{2} = \dfrac{a}{b}$ (where this fraction has been fully cancelled)

$\therefore \quad 2 = \dfrac{a^2}{b^2}$

$\therefore \quad 2b^2 = a^2$ since $2b^2$ is always even, then a must be even, e.g., $a = 2p$

$\therefore \quad 2b^2 = (2p)^2$ (replacing a by $2p$)

$\therefore \quad 2b^2 = 4p^2$

$\therefore \quad b^2 = 2p^2$ since $2p^2$ is always even, then b must be even.

We have proved both a and b are even, which means that they have a common factor of 2, so the fraction a/b cancels by 2. This is clearly a contradiction, therefore the original assumption that $\sqrt{2}$ could be expressed as a fraction must have been invalid.

isometric Having equal axes and lines. For

example, isometric graph paper is paper ruled into **equilateral triangles** rather than **squares**. See **isometry**.

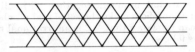

isometric   Isometric graph paper.

isometry   A **transformation** under which all lengths are **invariant**. For example, **translations**, **rotations**, **reflections** are isometries; **shears** and **enlargements** are not.

Isometries that preserve **sense** also are called *direct* isometries, and those that change sense are called *opposite* isometries. See diagram opposite.

isomorphic   Two groups are isomorphic if they have the same structure. There is a one–one

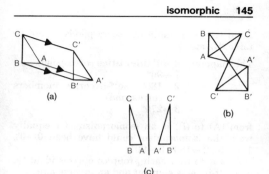

isometry (a) Translation (direct); (b) rotation (direct); (c) reflection (opposite).

correspondence φ between the groups such that φ(a∗b)=φ(a).φ(b) for all a and b. φ is called an *isomorphism*. For example:

(A)	0 1 2 3
0	0 1 2 3
1	1 2 3 0
2	2 3 0 1
3	3 0 1 2

(addition)
modulo 4

(B)	0° 90° 180° 270°
0°	0 90 180 270
90°	90 180 270 0
180°	180 270 0 90
270°	270 0 90 180

(combining)
rotations

Note: 0 in table (A) above, corresponds to 0° in

table (B). 1 in table (A) corresponds to 90° in table (B), etc.

The mapping: 0→0° (**identities** correspond)
　　　　　　 1→90°
　　　　　　 2→180° (**self-inverse** members correspond)
　　　　　　 3→270°

from (A) to (B) is an isomorphism. Or equally well, the isomorphism could have been 0→0°, 1→270°, 2→180°, 3→90°.

Also note that each group comprises identity, one self-inverse element and an inverse pair.

(C)	a b c d	(D)	1 2 3 4	(E)	1 2 4 3
a	a b c d	1	1 2 3 4	1	1 2 4 3
b	b a d c	2	2 4 1 3	2	2 4 3 1
c	c d a b	3	3 1 4 2	4	4 3 1 2
d	d c b a	4	4 3 2 1	3	3 1 2 4
			(multiplication)		
			modulo 5		

(C) is not isomorphic to (A) and (B) since every element of (C) is self-inverse. (D) is isomorphic to (A) and (B) but the table may need rearranging as in (E) to make this more obvious.

The set of **positive real numbers** under the operation of **multiplication** is isomorphic to the set of **real numbers** under **addition** using the isomorphism $x → \log x$. This is the basis for the use of the **logarithm function** as a calculating aid. The study of isomorphic systems is an important branch of modern **algebra**, as results proved about one system can be applied to the second.

isosceles An isosceles **triangle** is one that has two sides and two angles equal. An isosceles **trapezium** is one whose two non-**parallel** sides are equal. Each of the figures below has one line of **symmetry**.

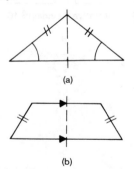

(a)

(b)

isosceles (a) An isosceles triangle and (b) an isosceles trapezium.

iterate   To repeat. So,

$$\frac{d^3y}{dx^3}$$

is an iterated differential, since y is **differentiated** three times. See **iterative**.

iterative   An iterative process involves the repetition of a sequence of operations to improve some result. See **iterate**.

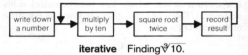

iterative   Finding $\sqrt[3]{10}$.

K

kilo- A prefix with **symbol** k which stands for one thousand (i.e. 10^3). For example, 1 kilobyte=1000 bytes.

kilogram A **unit** of mass: 1 kilogram=1000 **grams** (1 kg=1000 g). 1 kg is approximately 2.2 lbs.

kilometre A **unit** of distance: 1 kilometre=1000 **metres** (1 km=1000 m). 1 km is approximately $\frac{5}{8}$ of a **mile**.

kite A **quadrilateral** having two pairs of **adjacent** sides **equal**. As the diagrams overleaf show, a kite has one line of **symmetry**. This diagonal **bisects** the figure. The second diagonal is at right angles to the first but neither bisects the first nor the figure.

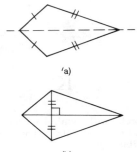

(a)

(b)

kite (a) Line of symmetry. The second diagonal is not a line of symmetry (b).

knot A **unit** of speed equivalent to one **nautical mile** per hour.

Königsberg bridge problem An historic problem in **topology** attributed to the famous Swiss mathematician Leonhard Euler (1707–83). It concerns the possibility of crossing each of the seven bridges in the town of Königsberg just once and returning to a starting point. By analysing the **network** of routes Euler showed the journey was not possible. See **Euler's formula**.

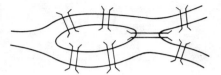

Königsberg bridge problem

L

Latin square   A square array of numbers or letters in which each number or letter appears once in every row and column. For example:

1 2 3 4			p q r s
4 3 2 1	but not		q r s p
3 1 4 2			r q p s
2 4 1 3			s p r p

The combination table for a **group** must be a Latin square.

latitude   A **circle** drawn on the earth's surface, whose **centre** is on the line joining the north and south poles. All latitude lines are **parallel** to each other, hence the term, 'parallel of latitude'. The latitude of any point is given by the **angle** ∠ AOB where O is the centre of the earth and A and B are points on the required latitude line and the **equator** respectively. A and B are on the same **longitude** line. All latitude lines, apart from the equator, are **small circles**. See **great circle**.

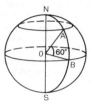

latitude   A has latitude 60° North.

limit   A **sequence** is said to tend to a limit if its terms approach a definite **value**, getting ever closer as more and more terms are included. Similarly the sum of a **series** may tend to a limit. For example:

$$1, 1\tfrac{1}{2}, 1\tfrac{3}{4}, 1\tfrac{7}{8}, 1\tfrac{15}{16} \ldots \text{limit is } 2$$
$$7.1, 7.01, 7.001, 7.0001 \ldots \text{limit is } 7$$
$$4+2+1+\tfrac{1}{2}+\tfrac{1}{4}+\tfrac{1}{8}+ \ldots \text{limit of the sum is } 8$$
$$6+2+\tfrac{2}{3}+\tfrac{2}{9}+\tfrac{2}{27}+ \ldots \text{limit of the sum is } 9$$
$$a+ar+ar^2+ar^3+ \ldots \text{limit of the sum of the}$$

geometric progression is
$$a/(1-r), \; (-1<r<1)$$

line of best fit   The line which fits a **set** of **data** most accurately when the data are plotted on a **graph**. In some cases this can be done sufficiently precisely by eye. For mathematical precision the *least squares method* is usually used. This involves finding how far above or below the proposed line each plotted point lies, squaring the distances, and adding the squares. The line of best fit is the line which produces the smallest total when the squares are added.

line graph   A **graph** constructed by joining a number of **points** together. These points represent known **values** of a given **variable**. The intermediate values indicated by **elements** of the **line segments** may or may not have a meaning.

line segment   A part of a straight line. In the illustration, AB is a line segment, being part of the line *l*. Points A and B do not necessarily have to be part of the line segment.

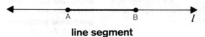

line segment

linear A general term for systems related in some way to straight lines.

A linear relationship between two **variables** x and y, is one that can be represented graphically by a straight line. **Equations** of such relationships (linear equations) can be written in the form $y=mx+c$.

The value of m gives the **gradient** of the **line graph**. The value of c gives the **intercept** of the graph on the y-**axis**.

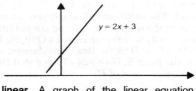

$y = 2x + 3$

linear A graph of the linear equation $y=2x+3$.

linear programming The process of finding optimum **values** of a linear **function** subject to limiting conditions or constraints. In practice, these functions often represent profits, volume of goods that can be produced, or production costs or times. A practical problem might involve dozens of **variables** and would then be solved using a **computer**. Simple problems with two variables can be solved using a **graph**.

For example, to maximize $U=3x+2y$ subject to the constraints

$$x+y \leqslant 13, \qquad 3x+y \leqslant 5$$
$$x \geqslant 0, \qquad y \geqslant 0$$

it may be seen that the constraints require the solution to lie in or on the boundary of the shaded region in the diagram below. Here the line AB represents $x+y=13$, and the line CD represents $3x+y=15$. The **solution** will always be one of the **vertices** of the shaded region, and the maximum value of $3x+2y$ will therefore be at O (0, 0), A (0, 13), E (1, 12) or D (5, 0). Brief calculations show that, at the point E, $U=3+24=27$ and that this is the maximum value of U.

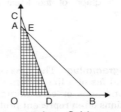

linear programming Solving a simple problem with a graph.

litre A measure of **volume** equal to 1000 **cubic centimetres**. 1 litre=1000 cm³. There are just over 4.5 litres in 1 **gallon**.

locus The path of a moving **point**. In the diagram below the locus of a point P, which moves so that it is equidistant from two fixed points A and B, is the perpendicular **bisector** of the line joining A and B.

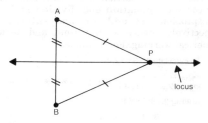

locus The locus of P is the perpendicular bisector of line AB.

logarithm Of a **number** to a given **base**, the **power** to which the base must be raised to give that number. For example:

$\log_{10} 100 = 2$ since $10^2 = 100$ (base=10)
$\log_2 64 = 6$ since $2^6 = 64$ (base=2)
$\log_5 0.008 = -3$ since $5^{-3} = 0.008$ (base=5)

Logarithms to base 10 are called **common logarithms** and to base **e** ($e = 2.718 \ldots$) are called **natural** or Naperian logarithms. The natural logarithm **function**, usually written $\log_e x$ or $\ln x$ is an important function in the study of **calculus**.

Before the introduction of calculators, logarithms were particularly useful since they reduced **multiplication** and **division** problems into problems of **addition** and **subtraction** respectively, hence saving time and tedious numerical working. For example:

To calculate 47.3×12.9

$\left.\begin{array}{l}\log_{10} 47.3 = 1.6749 \\ \log_{10} 12.9 = 1.1106\end{array}\right\}$ adding logs we get 2.7855

Antilog $2.7855 = 610.2$

Hence $47.3 \times 12.9 \approx 610.2$

longitude A semi-circle drawn on the surface of the earth joining the north and south poles. Longitude lines are parts of **great circles**, that is, circles having the same **radius** as that of the earth.

Just as **latitude** is measured in degrees north or south of the **equator**, so longitude is measured in degrees east or west of Greenwich, a fixed longitude line passing through Greenwich, London. (Known as the Greenwich meridian.)

If, in the diagram, NPS represents the Greenwich line (0°) and angles POA and POB are 30° and 20° respectively, then all points on the line NAS are on the longitude 30°W line, and all points on line NBS are on the 20°E line.

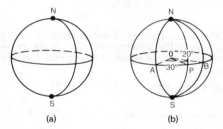

(a)   (b)

longitude (a) A longitude line. (b) All points on NAS are on the line 30°W, on NBS on the line 20°E.

lower bound Of a **set** of numbers, a number that is less than or equal to every **member** of the set. For example, $\{\frac{1}{2}, \frac{1}{3}, \frac{1}{4}, \frac{1}{5} \ldots\}$ has a lower bound of 0.

lowest **1.** The *lowest common multiple* (LCM) of two or more numbers is the smallest number that they will divide into exactly. For example:

LCM of 2, 3, 4 is 12
LCM of 18, 27 is 54
LCM of 8, 16 is 16
LCM of x, $2x$, x^2 is $2x^2$

2. The *lowest common* **denominator** of a set of fractions is the denominator that is the lowest common multiple of all the denominators. For example:

lowest common denominator of $\frac{3}{4}$, $\frac{5}{6}$ and $\frac{1}{3}$ is 12

lowest common denominator of $\dfrac{2}{x+1}$ and $\dfrac{4}{x^2-1}$ is x^2-1

3. A **fraction** is said to be in its lowest terms if it has been **cancelled** as much as possible. For example:

$\dfrac{3}{4}$, $\dfrac{5}{7}$, $\dfrac{1}{x-1}$, $\dfrac{3}{x^2}$ are in their lowest terms,

but

$\dfrac{6}{9}$, $\dfrac{20}{25}$, $\dfrac{x-1}{x^2-1}$, $\dfrac{3x}{x^4}$

are not since they will cancel down to

$\dfrac{2}{3}$, $\dfrac{4}{5}$, $\dfrac{1}{x+1}$ and $\dfrac{3}{x^3}$ respectively.

M

magic square A square array of numbers in which every **row**, **column** and the two **diagonals** add up to give the same total.

8	1	6
3	5	7
4	9	2

(a)

27	1	17	7
16	8	26	2
5	15	3	29
4	28	6	14

(b)

magic square   (a) Total 15; (b) total 52.

Example (b) illustrated is rather special since in addition each block of four corners adds up to 52, the block of four in the centre adds up to 52 and the shorter diagonals $(16+1+6+29)$ and $(5+8+17+2)$ add up to 52. The four corners add up to 52 and the middle two numbers of the top and bottom rows add up to 52. Also the middle two numbers of the left and right columns add up to 52!

magnitude The length of a **vector**, when represented in **line segment** form.

$\begin{pmatrix} 3 \\ 4 \end{pmatrix}$

magnitude By **Pythagoras' theorem** $3^2 + 4^2 = 5^2$. So the magnitude is 5.

major 1. The longer of the two axes of **symmetry** of an **ellipse** is called the major **axis**. The other is called the **minor** axis. For an ellipse with **equation**:

$$\frac{x^2}{a^2} + \frac{y^2}{b^2} = 1 \, (a > b)$$

the major **axis** has length $2a$.

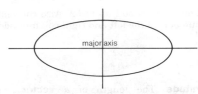

major The larger axis is the major axis

2. The shaded part of the diagram below shows a major **sector** of the **circle**. The arc ACB is called a major **arc** of the circle. See **minor**.

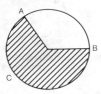

major A major sector of a circle. ACB is a major arc.

mapping A **function**. If the **members** of the **set** {8, 6, 4, 2} are each halved, a new set {4, 3, 2, 1} is formed. Halving transforms members of one set into members of another.

Diagram (a) below illustrates what happens to the numbers 8, 6, 4, 2. We say that the members of one set are mapped onto the members of the other. The numbers 4, 3, 2, 1 are called the **images** of 8, 6, 4, 2.

A mapping or function is a special type of relation in which every member of the set has one image.

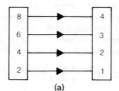

(a)

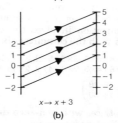

$x \rightarrow x + 3$

(b)

mapping (a) Transforming members of one set into members of another. (b) A mapping diagram.

matrix A rectangular array of numbers. For example:

$$A = \begin{pmatrix} 3 & 5 \\ 2 & 4 \end{pmatrix} \qquad B = \begin{pmatrix} 6 & 3 & 1 \\ 5 & 2 & 7 \end{pmatrix}$$

$$C = \begin{pmatrix} 8 & 1 \\ 6 & 2 \\ 5 & 3 \end{pmatrix} \qquad D = \begin{pmatrix} 10 & 5 \\ 4 & 2 \end{pmatrix}$$

Matrix A has two **rows** and two **columns**. Matrix B has two rows and three columns.

Matrices of the same order (i.e. with the same number of rows and columns) can be added, by adding corresponding elements. For example:

$$A + D = \begin{pmatrix} 3+10, & 5+5 \\ 2+ \ 4, & 4+2 \end{pmatrix} = \begin{pmatrix} 13 & 10 \\ 6 & 6 \end{pmatrix}$$

Matrices A and B are compatible for multiplication because A has as many columns as B has rows. Multiplication involves the combining of the rows of A with the columns of B:

$$A \times B = \begin{pmatrix} 3\times6+5\times5, & 3\times3+5\times2, & 3\times1+5\times7 \\ 2\times6+4\times5, & 2\times3+4\times2, & 2\times1+4\times7 \end{pmatrix}$$

$$= \begin{pmatrix} 43 & 19 & 38 \\ 32 & 14 & 30 \end{pmatrix}$$

Square matrices with non-**zero determinants** have multiplicative **inverses**:

The inverse of $\begin{pmatrix} a & b \\ c & d \end{pmatrix} = \left(\dfrac{1}{ad-bc} \right) \begin{pmatrix} d & -b \\ -c & a \end{pmatrix}$

For example: $A^{-1} = \begin{pmatrix} 2 & -2\frac{1}{2} \\ -1 & 1\frac{1}{2} \end{pmatrix}$

Matrices are a very powerful tool in the study of the solutions of **linear** equations.

maximum point A point on a **graph** at its highest **value**.

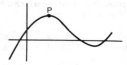

maximum The graph has a maximum point at the point P.

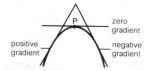

maximum A tangent drawn to the curve at P.

The **tangent** drawn to the curve at point P would have zero **gradient**. Just to the left of P, the tangent would have a positive gradient and just to the right it would be negative. See **minimum**.

mean   The arithmetic mean of a **set** of numbers is the **sum** of all the numbers divided by the number of figures in the set. (Often called the **average**). For example, the mean of 2, 6, 8, 9, 12 is:

$$\frac{2+6+8+9+12}{5} = \frac{37}{5} = 7.4.$$

In a dice game, the mean score can be calculated as follows. The scores are:

score on a dice	1	2	3	4	5	6	(total 20
number of throws	5	3	4	2	3	3	throws)

The mean score is:

$$\frac{(1\times5)+(2\times3)+(3\times4)+(4\times2)+(5\times3)+(6\times3)}{20}$$

$$=\frac{64}{20}=3.2.$$

mean deviation   More strictly speaking, this is the mean **absolute** deviation from the arithmetic **mean** of a list of numbers, defined as:

$$\frac{\text{sum of all deviations from the mean}}{\text{number of figures in the list}}$$

For example, for 7, 3, 10, 4 the mean is $\frac{24}{4}=6$, and hence the deviations from the mean are 1, 3, 4, 2. Therefore the mean deviation is:

$$\frac{1+3+4+2}{4}=\frac{10}{4}=2\tfrac{1}{2}.$$

measures of dispersion Dispersion measures the extent to which a **random** variable or set of observations is spread about its **average**. Two different **distributions** may have different dispersions, although they may have the same **mean**. In the diagram below distribution A has greater dispersion than distribution B, although they share the same mean.

There are various measures of dispersion: **mean deviation**, **standard deviation** and **semi-interquartile range** measure how far above or below the middle a typical member might be found. The others are **range** and **interquartile range**.

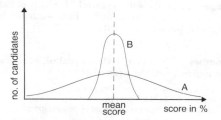

measures of dispersion Distribution A has a greater dispersion than B, although they share the same mean.

median The value of the middle number of a **set** of numbers, when they are arranged in ascending order. For example:

For 1, 2, 4, 7, 9 the median is 4.
2, 5, 8, 3, 1, 7, 6 becomes 1, 2, 3, 5, 6, 7, 8
    and the median is 5.

If there is no middle number, the average of the two middle numbers is taken. For example:

$$\text{For 1, 5, 7, 8, 9, 10 the median is } \frac{7+8}{2}=7\tfrac{1}{2}.$$

The median of a **frequency distribution** is found from the **cumulative frequency** graph or **ogive**.

mega- A prefix with **symbol** M which stands for one million (i.e. 10^6). For example, 1 megabyte=one million bytes.

member A number, letter, **symbol**, etc., that belongs to a **set**. For example:

  2 is a member of the set of even numbers
  a is a member of the set of vowels
  but 4 is *not* a member of the set of odd numbers

The mathematical symbols used for 'is a member of' and 'is not a member of' are ∈ and ∉ respectively. For example:

2 **∈** {prime numbers}

s **∈** {letters of the alphabet}

p **∉** {vowels}

α **∉** {β, γ, π, ϱ}

mensuration The measuring of geometrical quantities, for example, lengths, **areas** and **volumes**.

meridian A **great circle** drawn on the earth's surface which passes through the north and south poles. For example, all **longitude** lines may be referred to as meridians.

meridian All longitude lines are meridians.

metre A **unit** of length (just over 39 **inches**.) 1 metre=100 **centimetres** (1 m=100 cm).

metric system A system of **units** of measurement in which the fundamental units of length, mass, etc., are divided and compounded by

factors of ten. For example, 1 **kilogram**=1000 **grams**, 1 **metre**=100 **centimetres**.

The prefixes **kilo-, hecto-, deca-, deci-, centi-, milli-**, denote multiples of 10^3, 10^2, 10, 1/10, 1/100, 1/1000.

mile A **unit** of length. 1 mile=5280 **feet** or 1760 **yards**. 1 mile is approximately 1.6 **kilometres**.

millimetre A **unit** of length. There are 10 millimetres in 1 **centimetre**. 1 millimetre equals 0.1 cm, 10 mm=1 cm.

million One thousand thousand. One million equals 10^6.

minimum A point on a graph at its lowest **value**. In (a) P is the minimum. The **tangent** drawn to the **curve** at that point would have zero **gradient**. Just to the right of P the gradient is positive and just to the left, it is negative.

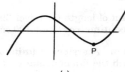

(a)

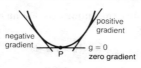

minimum (a) The minimum point is P. (b) A tangent drawn at P has a zero gradient.

minor 1. The shorter of the two axes of **symmetry** of an **ellipse** is called the minor **axis**. The

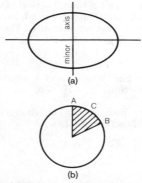

minor (a) The minor axis of an ellipse; (b) a minor sector of a circle.

other is called the **major** axis. For an ellipse with **equation**:

$$\frac{x^2}{a^2} + \frac{y^2}{b^2} = 1 \; (a > b)$$

the minor axis has length $2b$.

2. The shaded part of the diagram, shows a minor **sector** of the **circle**. The **arc** ACB is called a minor arc of the circle.

See **major**.

minute   **1.** A **unit** of time. There are 60 **seconds** in 1 minute, and 60 minutes in 1 **hour**.

2. A unit of angular measure. There are 60 minutes in 1 **degree** ($60' = 1°$).

mixed number   The **sum** of a **whole number** and a **fraction**: $2\frac{1}{2}$, $3\frac{2}{5}$, $-6\frac{7}{8}$ are all mixed numbers.

A mixed number can be changed into an **improper fraction** as follows:

$$2\frac{5}{7} = \frac{(2 \times 7) + 5}{7} = \frac{19}{7} \text{ or } 3\frac{9}{10} = \frac{(3 \times 10) + 9}{10} = \frac{39}{10}$$

mode   The number that occurs most often in a **set** of numbers. For example:

1, 2, 3, 3, 4, 6, 9: mode is 3
3, 4, 4, 4, 7, 7, 8: mode is 4
2, 2, 3, 5, 6, 9, 9: this set has two modes, 2 and 9,
and is said to be **biomodal**.

In the following dice game score

score on die	1	2	3	4	5	6
number of throws	5	6	2	4	3	5

mode is 2, since 2 was thrown most often (6 times)

In a grouped **frequency** distribution the group that contains most members is called the modal group or modal class. For example:

height of children (cm)	50–54	55–59	60–64	65–69
number of children	7	2	4	3

The modal group is 50–54 cm.

modulo arithmetic Refers to all numbers represented by their remainder when divided by n. In modulo 8, $6+5=3$, $4\times3=4$, etc.

This type of arithmetic can be modelled by moving round a clockface, as illustrated overleaf.

In arithmetic modulo p, where p is **prime**, equations of the form $ax+b=c$ have unique **solutions**.

modulo arithmetic   6+5≡start at 6, then move on 5 places, to finish at 3.

modulus 1. The **absolute value** of a **real number** x, denoted by $|x|$ is the positive value of x, regardless of its sign. For example: $|3.5|=3.5$, $|-7|=7$.

2. The distance of a **complex number** from the **origin** when represented on the **Argand diagram**.

The modulus of the complex number $a+bi$, denoted by $|a+bi|$, is $\sqrt{a^2+b^2}$.

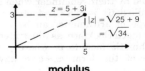

$$z = 5 + 3i$$
$$|z| = \sqrt{25 + 9}$$
$$= \sqrt{34}.$$

modulus

multiple A number or expression that can be divided by a given number or expression. The number x is a multiple of the number y if y divides x exactly. For example, 5, 10, 15, 20 ... are all multiples of 5.

multiplication One of the fundamental operations of **arithmetic**. Multiplication is associated with repeated addition. For example,

$$5\times3=5+5+5=15$$

Analogous operations on abstract mathematical systems are sometimes known as multiplication. For example:

matrix multiplication $\begin{pmatrix} 3 & 1 \\ 4 & 2 \end{pmatrix} \times \begin{pmatrix} 1 & 2 \\ 3 & 4 \end{pmatrix}$

$$= \begin{pmatrix} 6 & 10 \\ 10 & 16 \end{pmatrix}$$

mutually exclusive events Events that cannot occur in one **outcome**. Where two **events** are mutually exclusive either one or the other may occur, but not both. Hence the **probability** that one or the other of two mutually exclusive events occurs is the **sum** of the probabilities of the separate events.

$P(A\cup B)=P(A)+P(B)$ where A and B are mutually exclusive events.

The reason for this is best seen from a **Venn diagram**.

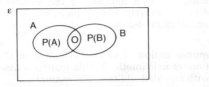

mutually exclusive event Since $A \cap B$ is empty, i.e. has probability 0, the result is clear.

N

natural numbers The set of **positive whole numbers** (counting numbers). The **set** of natural numbers $\{1, 2, 3, 4, \ldots\}$ is often denoted by $\mathbb{N}$.

natural logarithms Logarithms to the **base** of e.

nautical mile (or **sea mile**) A distance of 6080 feet, equivalent to one **minute** of **arc** measured along a **great circle** on the earth's surface. Hence 60 nm of great circle **subtends** an **angle** of 1° at the **centre** of the earth.

negative Describes a number whose value is less than **zero**. For example, -2.5, -724, -0.176, $-\frac{1}{10}$ are all negative; 0, 3.5, 176, 0.001 are not negative. See **positive**.

net A surface which can be folded into a **solid**. A shape is said to form a net for a, say, a **cube**, if they were cut out and could be folded up to form a cube.

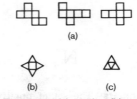

(a)

(b)   (c)

net The nets of (a) a cube, (b) a square-based **pyramid** and (c) a **tetrahedron**.

network A system of **points** (called **nodes**) linked together by **arcs**. A network divides the **plane** into a number of **regions**. The numbers of nodes, arcs and regions are related by **Euler's formula**:

nodes+regions=arcs+2

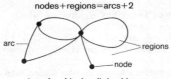

arc

regions

node

network Nodes linked by arcs.

Newton A **unit** of force. A resultant force of 1 Newton acting on a body of mass 1 kg produces an **acceleration** of 1 **metre** per **second** per second.

(Named after the English mathematician and scientist, Isaac Newton (1642–1727), who made outstanding contributions to the study of calculus, mechanics and optics.)

Newton's method A step-by-step, **iterative** method of finding the **roots** of an **equation** of the form: $f(x)=0$. It is illustrated in the **graph**.

If p_1, is an approximation to the root, then p_2, the **intercept** of the **tangent** to f at $x=p_1$ with the x-axis, is a better approximation.

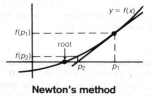

Newton's method

node A **point** on a **network** to which one or more **arcs** lead.

node

nonagon A nine-sided **polygon**. A **regular** nonagon has all its sides **equal**, and each of its interior **angles** measures 140°.

normal Denoting any point on a **curve** at which the line is **perpendicular** to the **tangent** at that point.

A line (or **vector**) is said to be normal to a **plane** if it is perpendicular to all lines belonging to the plane. In the diagram opposite, l and m are lines belonging to the plane. AB is perpendicular to l and m and hence the plane.

normal distribution A continuous distribution of a **random** variable with its **mean**, **median** and **mode** equal. The **graph** of the **probability** density **function** of the normal distribution is a continuous bell-shaped curve, symmetrical about the mean (see diagram opposite).

This distribution is very important in statistics especially because it is the distribution of many real-world variables, such as the weights of all 15-year-olds in a town.

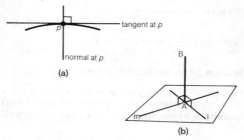

normal The normal to (a) a curve and (b) a plane.

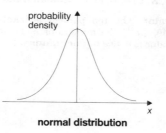

normal distribution

notation Symbols representing quantities, **operations**, **relations**, etc., together with the conventions regarding their use. For example:

The notation 5! means $5 \times 4 \times 3 \times 2 \times 1$

The notation $\displaystyle\sum_{1}^{4} i$ means $1+2+3+4$

nought The **numeral** 0 or **zero**.

null set The **empty set** denoted by $\varnothing$. It is a **set** with no **members**, for example, the set of **real numbers** whose **square** is **negative** in the null set.

numerals Symbols used to denote numbers. For example:
(a) 0, 1, 2, 3, 4, 5, 6, 7, 8, 9 are Arabic numerals.
(b) I, V, X, L, C, D, M are **Roman numerals**.

numerator The top part of a **fraction**. For example, for $\frac{17}{4}$ the numerator is 17, for $\frac{6}{13}$ the numerator is 6. See **denominator**.

O

observation sheet   A form designed to simplify the collection of **data**, perhaps where speed of recording is important.

For example, a traffic survey investigates the number of vehicles using a particular road during the morning rush hour. The observation sheet must enable the observer to record each bicycle, motorcycle, private car, van and lorry which passes, and the direction of travel of each (see diagram overleaf).

obtuse   An **angle** greater than 90° but less than 180°.

obtuse angle

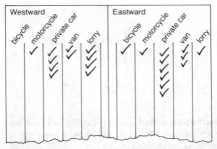

observation sheet A traffic survey form.

octagon A **polygon** having eight sides. A **regular** octagon has all its sides **equal** and all its interior **angles** are 135°.

octagon

octahedron A **polyhedron** having eight faces. A **rectangular** octahedron is made from eight **equilateral** triangles.

octahedron

odd number   A number that cannot be divided exactly by 2. For example, 7, 19, -107 are odd numbers.

   If n is a **whole number** then $2n+1$ is always odd.

ogive   Another name for a **cumulative frequency** curve.

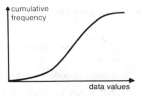

ogive   A typical shape.

operation   A way of combining **members** of a

set. Binary operations combine two members of a set to produce a third member as a result.

The study of abstract operations and their properties forms an important branch of modern **algebra**.

order   1. The number of **rows** and **columns** of a **matrix**.

$$A = \begin{pmatrix} 3 & 1 & 6 \\ 2 & 5 & 8 \end{pmatrix}$$ Matrix A has order 2×3. This means it has two rows and three columns.

2. The number of **arcs** that lead to a node in a **network**.

order   The order of nodes in a network.

3. A figure which maps to itself by **reflections** about n lines has line **symmetry** of order n.

A figure which maps to itself in n ways by rotations about some centre O has rotational symmetry of order n about O.

(a)   (b)

order   (a) Rotational symmetry of order 3;
(b) a square has line symmetry of order 4.

4. A **group** with n elements is said to be a group
of order n.

order of accuracy   A **solution** of a numerical
problem is of the right order of accuracy if its
digits are in the correct **decimal** columns. This
can usefully be checked by making an **estimate**
of the solution.

For example, since we may estimate that
$278 \div 39$ is about 7 it is clear that solutions to the
same problem such as 71.282 (3 dp) and 0.713
(3 dp) are of the wrong order of accuracy. 7.128
(3 dp) is of the right order of accuracy.

ordered pair   A pair of numbers whose **values**
and **order** are **significant**. Used in particular to
denote a pair of **Cartesian coordinates** specifying a **point** in a **plane**.

Ordered pairs such as (H, T), (H, H) etc., may also be used to specify the results of trials, for example, tossing two coins.

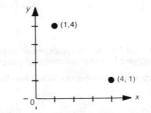

ordered pair The ordered pairs (1, 4) and (4, 1) specify different points.

ordinate The *y*-coordinate, or distance from the **horizontal axis**, of a point referred to a system of **Cartesian coordinates** (see diagram opposite).

origin The point where the *x*- and *y*-axes cross. It has **coordinates** (0, 0) and the coordinates of all other points are measured relative to the origin (see diagram opposite).

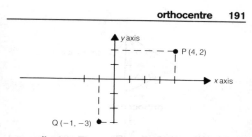

ordinate The ordinate of P is 2, of Q is −3.

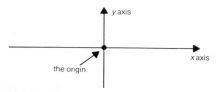

origin The origin on a Cartesian grid.

orthocentre The point where the **altitudes** meet in a **triangle** (see diagram overleaf).

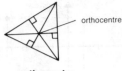

orthocentre

ounce   A **unit** of mass. There are 16 ounces in 1 **pound**. (16 oz=1 lb) and just over 28 **grams** in 1 ounce.

outcome   The result of a statistical experiment or other activity involving uncertainty.

P

parabola The **locus** of a **point** which moves so that it is equidistant from a fixed point (called the **focus**) and a fixed line (the **directrix**). It is a **curve** of the **conic** family.

Any equation of the form $y = ax^2 + bx + c$ or $y^2 = 4ax$ will give a parabolic shape. The path of a **projectile** is approximately parabolic.

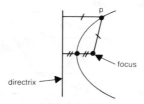

parabola $y^2 = 4ax$ has a focus at point $(a, 0)$ and directrix the line $x = -a$.

parallel Lines in a **plane** that never meet no matter how far they are extended. Lines are always the same distance apart. A pair of railway lines are parallel.

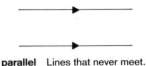

parallel Lines that never meet.

parallelepiped A **polyhedron** whose **faces** are all **parallelograms**.

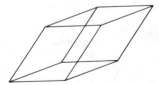

parallelepiped

parallelogram A **quadrilateral** with its opposite sides **equal** and **parallel**.

All parallelograms have the following properties:

(a) Diagonals **bisect** each other, and the whole shape
(b) Opposite **angles** are equal
(c) Rotational **symmetry** of **order** 2.

parallelogram AB=DC and AB||DC; AD=BC and AD||BC.

parameter A quantity that, when varied, affects the **value** of another. In the **equation** $y=mx+c$, the values m and c are called parameters and specify the characteristics of the straight line represented by the equation. (m represents the **gradient**, and c the **intercept** on the y-axis.)

The parametric equation of the **parabola** $y^2=4x$ is $x=t^2$, $y=2t$, each value of the parameter t gives a point on the **curve**.

Pascal's triangle A triangular pattern of numbers. Each number is the **sum** of the two numbers directly above it. The numbers in each **row** of the **triangle** are the **binomial coefficients** which occur in the expansion of $(x+y)^n$ for various values of n.

```
            1
          1   1
        1   2   1
      1   3   3   1
    1   4   6   4   1
  1   5  10  10   5   1
```

Pascal's triangle

penny A **unit** of currency, one hundred pennies (pence)=one **pound** (100p=£1).

pentagon A five-sided **polygon**. A **regular** pentagon has all its **sides equal** and each of its interior **angles** is 108°.

pentagon

percentage A method of relating a **fraction** of a given quantity to the whole in parts per hundred. For example:

30% of something is thirty parts per hundred.

Hence $30\% \equiv \dfrac{30}{100}$ (note % means 'per cent')

Percentage changes in a quantity are calculated as follows:

If original value=A
new value=B
per cent change $= \dfrac{|A-B|}{A} \times 100$

percentile The **range** of a **frequency distribution** is divided into 100 parts by percentiles. 15% of the data lies below the fifteenth percentile, 40% lies below the fortieth percentile, etc.

The **median** value of a distribution occurs at the fiftieth percentile.

perfect number A number that is **equal** to the **sum** of its **factors** (excluding the number itself). For example:

6=1+2+3   28=1+2+4+7+14

The third perfect number is 496.

perimeter The distance measured round the **boundary** of the figure. The perimeter of a circle is also known as the **circumference**.

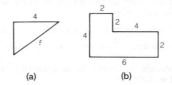

(a)

(b)

perimeter The perimeter of triangle (a) is 3+4+5=12 units, of the shape (b) is 2+2+4+2+6+4=20 units.

periodic function A regularly repeated **function**. The **graph** of $y=\sin x$ repeats itself every 360°. This value 360° is said to be the *period* of the function and $y=\sin x$ is said to be a periodic function (see diagram opposite).

permutation An ordered arrangement of a **set** of numbers.

For example, all possible permutations of the numbers 1, 2, 3 are 1, 2, 3, 12, 13, 23, 21, 31, 32, 123, 132, 231, 213, 312 and 321.

There are $n!$ (**factorial** n) permutations of n numbers taken all at a time. For example:

       1 number has 1 permutation
       2 numbers have 2 permutations

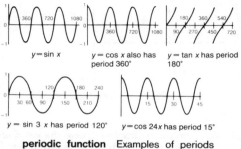

$y = \sin x$

$y = \cos x$ also has period 360°

$y = \tan x$ has period 180°

$y = \sin 3x$ has period 120°

$y = \cos 24x$ has period 15°

periodic function   Examples of periods of various functions.

3 numbers have 6 permutations
4 numbers have 24 permutations

There are $n!/(n-r)!$ permutations of n numbers taken r at a time.

perpendicular   At **right angles**. Two straight lines are said to be perpendicular if they meet at right angles.

Two planes may also be perpendicular. A line which is perpendicular to a plane is said to be **normal** to the **plane** (see diagram overleaf).

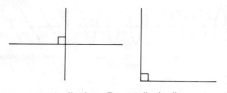

perpendicular   Perpendicular lines.

pi   The **symbol** (π) used to denote the **ratio** of the **circumference** of a **circle** to its **diameter**.

$$\pi = \frac{C}{D}$$

π is an **irrational** number: $\pi = 3.1415926535 \ldots$ We commonly use 3.14 or 22/7 as an (approximate) value for π.

π is used in calculating lengths, **areas** and **volumes** of circular figures and **solids**. For example:
(a) The area of a **circle** of **radius** R is πR^2
(b) The volume of a **cylinder** is $\pi R^2 h$
(c) The volume of a **sphere** is $\frac{4}{3} \times \pi R^3$

pictogram   A way of representing information (usually statistical) in the form of a picture.

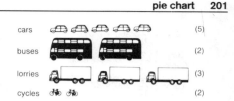

cars		(5)
buses		(2)
lorries		(3)
cycles		(2)

pictogram Pictogram showing vehicles passing a checkpoint in one minute.

pie chart A circular diagram which is another way of representing statistical information. The **circle** is divided into **sectors** whose areas represent the number in the **set**.

pie chart The numbers of cars, lorries, buses and cycles making up the set 'vehicles'.

For example, for the information in the diagram above, the angles would be:

$$\text{Car} \quad \frac{5}{12} \times 360° = 150°$$

$$\left.\begin{array}{l}\text{Bus}\\\text{Cycle}\end{array}\right\} \quad \frac{2}{12} \times 360° = 60° \text{ each}$$

$$\text{Lorry} \quad \frac{3}{12} \times 360° = 90°$$

pint A measure of **volume**. There are 8 pints in 1 **gallon** and approximately 1.76 pints in 1 **litre**.

plan The view of a shape when looking vertically downwards onto it. The plan view of a cylinder is a circle. A map gives a plan view of a portion of country.

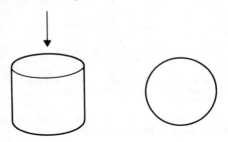

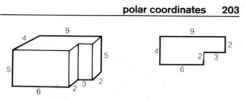

plan Objects and their plan views.

plane A flat surface. In **Cartesian coordinates** the **equation** of a plane is of the form $ax+by+cz=d$.

point A location in space or on a surface. Points are often described by their **coordinates**. A point has position but no real size.

polar coordinates A system of specifying the position of **points** by their distance r from a fixed point, and angle θ measure from a fixed line.

Polar coordinates are of the form (r, θ) as opposed to (x, y) of **Cartesian coordinates** (see diagram overleaf).

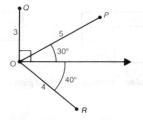

polar coordinates *P* is (5, 30°), *Q* is (3, 90°) and *R* is (4, 320°) or (4, −40°).

polygon A **plane** shape having three or more sides. In general a polygon having *n* sides will have *n* interior **angles** that sum to (*n*−2)×180°.

  Regular polygons have all their sides and all their angles **equal** (see diagram opposite).

polyhedron A **solid** shape each of whose **faces** is a **polygon** (see diagram opposite). There are only five **regular** polyhedra: **tetrahedron, cube, octahedron, dodecahedron, icosahedron.**

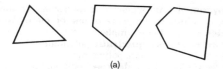

(a)

Name of polygon	Number of sides	Sum of interior angles
Triangle	3	180°
Quadrilateral	4	360°
Pentagon	5	540°
Hexagon	6	720°
Heptagon	7	900°
Octagon	8	1080°
Decagon	10	1440°
Duodecagon	12	1800°
Icosogon	20	3240°

(b)

polygon (a) Examples of polygon shapes; (b) polygons and their attributes.

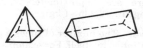

polyhedron Each face of the solids is a polygon.

polynomial An algebraic expression containing only **positive** powers of one or more **variables** x, y, ... For example:

(a) $3x^2+5x+1$ is a **quadratic** polynomial.

(b) y^3+7y+2 is a *cubic* polynomial.

(c) $2x^2y^3+3xy+1$ is a **degree** 5 polynomial.

In the polynomial $5x^4+3x^2+7x-6$, the number 5 (which multiplies x^4) is called the **coefficient** of x^4.

population A statistical term referring to the **set** of items for which a certain characteristic is being measured. Populations may be finite or **infinite**.

positive Describes a number whose **value** is greater than **zero**. For example, 735, $3\frac{1}{4}$, 0.1 are all positive; 0, -2, -3.5 are not positive. See **negative**.

pound 1. A **unit** of currency. £1 equals 100 pence.

2. A unit of weight. One **pound** equals 16 **ounces** (1 lb=16 oz), 14 pounds equals 1 stone. There are approximately 2.2 pounds in 1 **kilogram** (2.2 lb≈1 kg).

power The number of times a quantity is to be multiplied by itself. For example 2 to the power 4 (written 2^4)=$2\times2\times2\times2$=16; 3 to the power 2 (written 3^2)=3×3=9.

Any number to the power 2 is said to be 'squared', any number to the power 3 is said to be 'cubed'.

prime number A number that has exactly two **factors**, 1 and the number itself. There are an **infinite** number of prime numbers. For example, 2, 3, 5, 7, 11, 13, 17, 19, etc. Note: 2 is the only **even** prime number.

prism A **polyhedron** having the same **cross-section** throughout its length. The **volume** of a prism is found by multiplying the cross-sectional area by the length of the solid.

prism A triangular prism.

probability The measure of how likely an **event** is. Its **value** lies between 0 (impossible event) and 1 (certain event). For example:
(a) Probability of the sun rising in the east=1.
(b) Probability that Sunday is the day after Monday=0
(c) Probability of throwing a 'head' with a coin=0.5

In situations where several equally likely outcomes are possible, the probability of a particular event can be measured by:

number of events favourable to the outcome
total number of possible events

See **equally likely event**.

product   The result of **multiplying** two or more numbers together. For example:
(a) Product of 2, 3, 4 is $2 \times 3 \times 4 = 24$.
(b) Product of 6 and 11 is $6 \times 11 = 66$.
The product of the matrices

$$\begin{pmatrix} 2 & 4 \\ 4 & 1 \end{pmatrix} \text{ and } \begin{pmatrix} 3 & 5 \\ 1 & 2 \end{pmatrix} \text{ is } \begin{pmatrix} 10 & 18 \\ 13 & 22 \end{pmatrix}$$

progression   A **sequence** of numbers each of which is related to the previous one by adding or multiplying by a fixed number. For example:
(a) 2, 5, 8, 11 is an **arithmetic** progression (add 3 to get the next term).
(b) 2, 6, 18, 54 is a **geometric** progression (multiply by 3 to get the next term).

projectile   A particle thrown or projected into the air. The path of a projectile, its **trajectory**, is roughly **parabolic**.

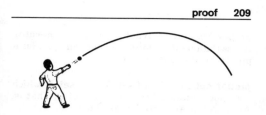

projectile   A projectile and its path.

projection   A **transformation** of a line or
shape into another. In the diagram below, if a
light was shone vertically downwards onto the
horizontal plane PQRS, the line AB would cast
its shadow along the line CD. CD is said to be the
projection of AB onto the plane.

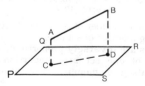

projection   CD is the projection of AB
onto the plane.

proof   The logical argument used to establish
the truth of a statement. See **theorem**.

proper fraction A **fraction** whose **numerator** is less than its **denominator**. For example, $\frac{3}{4}$ is a proper fraction but $\frac{7}{3}$ is not.

proper subset A **subset** A of a **set** B which does not contain all the **elements** of B. That is, $A \cap B = A$ but $A \cap B \neq B$. For example:

$\{1, 2, 3\}$ is a proper subset of $\{1, 2, 3, 4, 5, 6\}$ but
$\{1, 2, 3, 4, 5, 6\}$ is not a proper subset of $\{1, 2, 3, 4, 5, 6\}$.

proportional Varying in a **constant** ratio to another quantity. Two sets of numbers are said to be in proportion when the **ratio** between corresponding members of the two **sets** is constant. For example:

$\{1, 2, 5, 8, 10\}$ and $\{2, 4, 10, 16, 20\}$ are in proportion, since $1{:}2 = 2{:}4 = 5{:}10 = 8{:}16 = 10{:}20$.

That is, each member of set two is twice the corresponding member of set one.

Two sets are **inversely** proportional when the members of one set are proportional to the **reciprocals** of the members of the other set. For example:

$\{1, 2, 3, 4, 6\}$ is inversely proportional to $\{12, 6, 4, 3, 2\}$

In this case the product of the corresponding elements is constant, i.e.

$$1 \times 12 = 2 \times 6 = 3 \times 4 = 4 \times 3 = 6 \times 2$$

The **circumference** of a **circle** is proportional to its **diameter** (C∝D). The *constant of proportionality is* π. (C = πD).

protractor A device used for measuring **angles**.

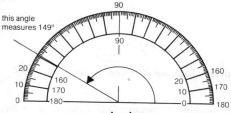

this angle measures 149°

protractor

pyramid A **polyhedron** with a **polygon** for its **base**, the other **faces** being **triangles** with a common **vertex**.

(a)   (b)   (c)

pyramid   (a) Square-based, (b) triangular and (c) hexagonal pyramids.

The volume of a pyramid is given by:

$V = \frac{1}{3}$ (area of base) × (perpendicular height)

$V = \frac{1}{3} Ah$

A **cone** is similar to a pyramid and the formula for its volume resembles that for a pyramid.

pyramid Volume of cone or pyramid: $V = \frac{1}{3}\pi r^2 h$.

Pythagoras' theorem This states that in any **right-angled** triangle, the area of the square on the **hypotenuse** is equal to the sum of the areas of the squares on the other sides.

For example, the following triples of numbers all correspond to the lengths of sides in right-angled **triangles**:

$$3, 4, 5 \ (3^2 + 4^2 = 9 + 16 = 25 = 5^2)$$
$$6, 8, 10$$
$$5, 12, 13$$
$$7, 24, 25$$
$$1, 1, \sqrt{2}$$
$$1, 2, \sqrt{5}$$
$$8, 15, 17$$
$$20, 21, 29, \text{ etc.}$$

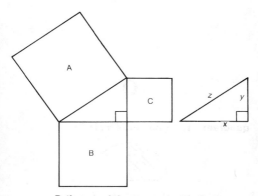

Pythagoras' theorem   A=B+C or
$z^2 = x^2 + y^2$.

Pythagoras' theorem is frequently used to find
unknown lengths in geometrical configurations.
(Named after the Greek philosopher and
mathematician Pythagoras (c. 580–500 B.C.)

Q

quadrant 1. A **quarter** of a **circle**.

quadrant The quadrant of a circle.

2. One of the four parts that the **plane** is divided into by the x- and y-axes:

	y-axis	
second quadrant		first quadrant
		x-axis
third quadrant		fourth quadrant

quadrant The quadrants of a plane.

quadratic equation A **polynomial** equation containing powers of x up to x^2. For example, $3x^2+5x=11$.

A quadratic equation can be written in the form

$$ax^2+bx+c=0 \text{ with } a\neq0.$$

Such equations may have two distinct, two coincident, or 0 real **solutions** depending upon the value of the **discriminant** of the equation.

The solutions of a general quadratic equation $ax^2+bx+c=0$ can be found by using the **formula**:

$$x=\frac{-b\pm\sqrt{b^2-4ac}}{2a}$$

The **graph** of quadratic **function** $y=ax^2+bx+c$ is **parabola**-shaped. See **quartic equation**, **quintic equation**.

quadrilateral A **polygon** having four sides. **Square**, **rectangle**, **rhombus**, **parallelogram**, **kite**, **trapezium** are all special kinds of quadrilaterals.

quadrilaterals Examples of quadrilaterals.

quarter The name given to one fourth part of a shape or quantity.

One quarter of 12 is 3, since $12 \div 4 = 3$ or $\frac{1}{4} \times 12 = 3$. See **quadrant**.

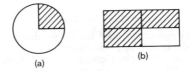

(a)   (b)

quarter One-quarter ($\frac{1}{4}$) of the circle (a) and three quarters ($\frac{3}{4}$) of the shape (b) have been shaded.

quartic A **polynomial** equation containing powers of x up to x^4. For example,

$$x^4 + 7x^3 - 2x^2 - 2x + 3 = 0.$$

See **quadratic equation, quintic equation**.

quartile One of the three points that divide a set of data into four equal parts. It is a statistical measure used in connection with **cumulative frequency**. The first or lower quartile of a **set** of data is the **value** below which one-quarter of the data lies. The second quartile (more commonly known as the **median**) is the value below which one half of the data lies, and the third or upper quartile is the value below which three-quarters of the data lies. They correspond to the 25th, 50th and 75th **percentiles**.

In a list of n numbers the lower quartile is the $\frac{1}{4}(n+1)$th number, the median is the $\frac{1}{2}(n+1)$th, and the upper quartile is the $\frac{3}{4}(n+1)$th. For example:

1, 2, 2, 5, 6, 7, 7, 7, 8, 9, 10

lower   median   upper
quartile   quartile

Lower quartile = $\frac{1}{4}(11+1)$th number = third value = 2
Median = $\frac{1}{2}(11+1)$th number = sixth value = 7
Upper quartile = $\frac{3}{4}(11+1)$th number = ninth value = 8

The **interquartile range** is 6.

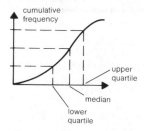

quartile Data are referred to according to how much lie below each point.

quintic equation A **polynomial** equation containing powers of x up to x^5. For example,

$$x^5 + 3x^4 + 2x^3 + 6 = 0.$$

See **quadratic equation, quartic equation**.

quotient The result obtained when a **division** is performed. For example:

$$21\overline{\smash{)}625}^{\textstyle 29(r16)}$$ The quotient is 29 and **remainder** 16.

$$
\begin{array}{r}
x+4 \\
x+1\overline{\smash{)}x^2+5x+4} \\
\underline{x^2+x} \\
4x+4 \\
\underline{4x+4} \\
0
\end{array}
$$

The quotient is $x+4$ with no remainder.

R

radian The **angle** that is **subtended** at the **centre** of a **circle** by a **minor arc** of length equal to the **radius** of the circle.

One radian≈57° and in a full circle there are 2π or 6.28 (approx.) radians. One d**e**gree equals $\frac{\pi}{180}$ radians.

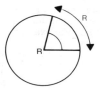

radian The angle subtended is 1 radian.

radius The distance from the **centre** of a **circle** to any point on its **circumference**.

random Based on chance. A **sample** taken from a **population** is said to be random if every member of the population has an **equal** chance of being chosen.

A random number is a number is which each **digit** is chosen from the **set** of values 0, 1, 2, 3, 4, ... 9, where each **value** has the same chance of being chosen.

range 1. A statistical term referring to the difference between the smallest and largest **members** of a **set** of numbers. For example, for 2, 3, 4, 7, 9, 10, 12, 15, the range is $15-2=13$.
2. The range of a **function** is the set of **images** of the members of the **domain** of the function. For example, for the function $x \rightarrow 3x+1$ with domain $\{1, 2, 3, 4\}$, the range is $\{4, 7, 10, 13\}$.

rank In a **set** of numbers or quantities arranged in ascending order, the position of any element in that list. For example:

For the numbers	4	8	5	6	2	6	0	12
Rank order is	0	2	4	5	6	6	8	12
	↓	↓	↓	↓	↓	↓	↓	↓
Rank	1	2	3	4	5½	5½	7	8

ratio One number divided by another. It is used to compare two quantities measured in similar units by considering the **quotient** of the two quantities. For example,

The ratio of £2 to 50p is written as 200:50=4:1.

Note: ratios are simplified similarly to **fractions**.

rational number Any number that can be expressed in the form of a **fraction**, i.e. in the form a/b where a and b are whole numbers, and $b \neq 0$. For example:

$$2\tfrac{1}{2} = \frac{5}{2} \qquad -7 = -\frac{7}{1} \qquad \sqrt{0.81} = 0.9 = \frac{9}{10}$$

The **symbol** for the **set** of all **rational** numbers is Q. Rational numbers can always be written as **terminating** or **recurring decimals**.

Numbers that cannot be written as a fraction are called **irrational** numbers.

rationalize In an algebraic expression to remove operations of the form $\sqrt{}, \sqrt[3]{}$, etc, without changing the value of the expression. For example,

$$\sqrt{x-2} = x \text{ rationalizes into } x-2 = x^2$$
or
$$x^2 - x + 2 = 0$$

$$\sqrt[3]{x^2+1} = 2 \text{ rationalizes into } x^2 + 1 = 8$$
or
$$x^2 - 7 = 0$$

In the example below the **denominator** of the fraction has been rationalized:

$$\frac{1}{\sqrt{x}+y} = \frac{\sqrt{x}-y}{x-y^2}$$

To obtain this result the **numerator** and denominator of the fraction are both multiplied by $\sqrt{x} - y$.

ray A straight line extending from a **point**, called the origin of the ray. Another name for a ray is a *half-line*.

ray Three rays all with 0 as common origin.

real numbers The set of numbers that is the **union** of the sets of **rational** and **irrational** numbers.

To each point on the continuous real number line there corresponds a real number. For example,

$$-3 \quad -2 \quad -1 \quad 0 \quad 1 \quad 2 \quad 3 \quad 4$$
$$-1.5 \quad 0.3 \quad \sqrt{2} \quad e \quad \pi$$

are all real numbers.

reciprocal The number 1 divided by a

quantity. The reciprocal of a non-zero number x is the number $1/x$. For example, the reciprocals of 5, $\frac{3}{4}$ and 0.4 are $\frac{1}{5}$, $1/\frac{3}{4}=\frac{4}{3}$, $1/0.4$, respectively.

The **product** of any number and its reciprocal is equal to 1.

rectangle A **quadrilateral**, all of whose **angles** are equal to 90°. Opposite pairs of sides are of equal lenth. A rectangle has two lines of **symmetry** and rotational symmetry of order 2. The **diagonals** of a rectangle **bisect** each other and the figure.

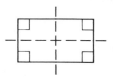

rectangle The lines of symmetry in a rectangle.

rectangle number A number having more than two **factors**. These numbers can be represented by **rectangular** configurations of dots. For example:

12=2×6	or 3×4

· · · · · ·   · · · ·
· · · · · ·   · · · ·
· · · ·

20=2×10   or   4×5
· · · · · · · · · ·   · · · · ·
· · · · · · · · · ·   · · · · ·
· · · · ·
· · · · ·

The set of rectangle numbers is: {4, 6, 8, 9, 10, 12, 14, 15, 16 ...}. The set of **prime numbers** is: {2, 3, 5, 7, 11 ...}. Notice that the **union** of these two disjoint sets gives the set of all **natural numbers** (apart from 1 which is unique in having just one factor).

rectangular   1. A rectangular **prism** is a prism whose **cross-section** is a rectangle (same as a **cuboid**).
2. The x and y **axes** are said to be rectangular axes, being perpendicular to each other.
3. A rectangular hyperbola is one whose **asymptotes** are the x and y axes. The equation of a rectangular hyperbola is $xy=c^2$.

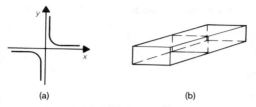

(a)   (b)

rectangular   (a) A rectangular prism; (b) a rectangular **hyperbola**.

rectilinear motion   Motion along a straight line.

recurring decimal   A **decimal** that contains an infinitely repeating block of decimal **digits**. For example,

$\frac{1}{6} = 0.16666\ldots$ written as 0.1̇6̇

$\frac{2}{11} = 0.181818\ldots$ written as 0.1̇8̇

All recurring decimals represent **rational numbers**. Fractions with **denominators** containing **prime factors** other than 2 or 5 will recur if written in decimal form. See **terminating decimal**.

re-entrant polygon   A **polygon** is said to be re-entrant if one or more of its angles is greater than 180° (i.e. is a **reflex** angle).

Non re-entrant polygons are called **convex**.

this angle is greater than 180°

re-entrant polygon

reflection A geometrical **transformation** of the **plane** in which **points** are mapped to **images** by folding along a *mirror line*.

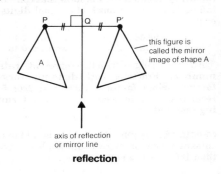

this figure is called the mirror image of shape A

axis of reflection or mirror line

reflection

The object and image are **congruent**. A line joining any point P to its image P′ will always be **perpendicular** to the mirror line and the distances PQ and P′Q will be equal (see diagram).

A **matrix** can be used to represent reflections in lines through the origin. Some of the simpler ones are below.

Reflection in:

$$x\text{-axis } (y=0) \qquad \begin{pmatrix} 1 & 0 \\ 0 & -1 \end{pmatrix}$$

$$y\text{-axis } (x=0) \qquad \begin{pmatrix} -1 & 0 \\ 0 & 1 \end{pmatrix}$$

$$\text{the line } y=x \qquad \begin{pmatrix} 0 & 1 \\ 1 & 0 \end{pmatrix}$$

$$\text{the line } y=-x \qquad \begin{pmatrix} 0 & -1 \\ -1 & 0 \end{pmatrix}$$

Note: points on the mirror line do not change their position after the reflection. See **mapping**.

reflex angle   An **angle** that is greater than 180° but less than 360° (see diagram overleaf).

reflexive relation   A **relation** on a **set** where every **member** of the set is related to itself.

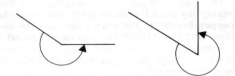

reflex angles

region The areas of a **plane** divided by a **network**. Regions are spaces enclosed by the **arcs** of the network.

The graph of a **linear** relation, for example, $x+y=5$, divides the plane into two regions.

Each region can be specified by an ordering. In this case $x+y<5$ and $x+y>5$.

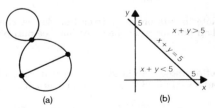

region (a) This network has four regions including that 'outside' it. (b) The line $x+y>5$ divides the plane into two regions.

regular Having all faces or sides of equal size and shape. A regular **polygon** is one having all its sides **equal**, and all its interior **angles** equal.

A regular **triangle** is called an **equilateral** triangle. A regular **quadrilateral** is a **square**.

A regular **polyhedron** is one in which all the **faces** are **congruent** regular polygons. There are just five regular polyhedra: **tetrahedron**, **cube**, **octahedron**, **dodecahedron**, **icosahedron**.

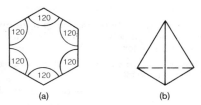

(a)   (b)

regular (a) A regular hexagon; (b) a regular tetrahedron made up of four equilateral triangles.

relation A connection between the **members** of a **set**, or between the members of two sets one to another. For example, 'is older than' on a set of people; 'is parallel to' on a set of lines.

Relations between two sets of numbers are of particular importance. For example:

$x \rightarrow x^2 + 1$ between the set of all real numbers and $[1, \infty]$.

remainder The number left when one number is divided by another. When 14 is divided by 4, the **quotient** is 3 (because $4 \times 3 = 12$) and there is a remainder of 2.

In any **division** process when the **dividend** is not a **multiple** of the **divisor**, there will be a remainder involved. For example:

$$362 \div 25 = 14, \text{ remainder } 12.$$

Note that $362 = (14 \times 25) + 12$, i.e.

dividend = (quotient × divisor) + remainder.

The remainder **theorem** for **polynomials** states that when a polynomial $P(x)$ is divided by $(x-a)$ the remainder is $P(a)$. For example:

when $P(x) = x^2 + 7x + 2$ is divided by $(x+3)$ the remainder is $P(-3) = (-3)^2 - (7 \times 3) + 2 = -10$.

Notice that if $(x-a)$ is a **factor** of $P(x)$, then $P(a) = 0$. For example:

$(x-2)$ is a factor of $P(x) = x^2 + 4x - 12$ and $P(2) = 0$

repeated root If a **root** of an **equation** occurs more than once it is said to be a repeated or multiple root of the equation. For example, the expression:

$$x^3 - 7x^2 + 16x - 12 = (x-2)\,(x-2)\,(x-3).$$

Hence the equation $x^3 - 7x^2 + 16x - 12 = 0$ has three roots 2, 2 and 3. 2 is a repeated root.

residue   When the numbers 8, 23, 38 etc. are divided by 5, they leave a **remainder** of 3, similarly 2, 17, 42, etc. leave a remainder of 2. We say that 8, 23, 38 all belong to the residue class $\overline{3}$ (**modulo** 5) and 2, 17, 42 belong to the residue class $\overline{2}$ (modulo 5). For modulo 5 there are five residue classes, denoted by $\overline{0}$, $\overline{1}$, $\overline{2}$, $\overline{3}$, $\overline{4}$ according to whether the remainder is 0, 1, 2, 3, 4.

resolve   To find the **component** of a **vector** any particular direction.

resolve   The resolved part of **d** in the x direction is **OB** and in the y direction **OA**.

In the diagram above, the resolved part of the vector **d** in the x direction is vector **OB** and in the y direction is **OA**. The **magnitudes** are

$$\begin{aligned}|\mathbf{OB}| &= |\mathbf{d}| \times \cos 40° \\ |\mathbf{OA}| &= |\mathbf{d}| \times \sin 40°\end{aligned}$$

resolve   The resolved part has magnitude
|**r**|×cos θ.

In the second diagram the resolved part of the
vector **r** in the direction **AB** has magnitude
|**r**|×cos θ where θ is the angle between the direc-
tions of **r** and **AB**.

resultant   Of two or more **vectors**, the single
vector that would have the same effect as all the
others combined. The resultant is obtained by
adding the **component** vectors.

resultant   Resultant is 10 **newtons** at an
angle of 36.9° to the force of 8 newtons.

revolve   To rotate about an **axis** or **point**.
If the shapes in the diagram are rotated 360°
about the axes marked, **solids** of revolution will
be formed. The triangle will produce a **cone** and

the semi-circle will produce a **sphere**. See **rotation**.

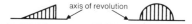

axis of revolution

revolve   The triangle would form a **cone** and the semicircle a sphere if rotated 360° about the axes.

rhombus   A **parallelogram** with all of its sides equal. A rhombus has two lines of **symmetry** and rotational symmetry of **order** 2. The diagonals of a rhombus **bisect** each other at right angles, and bisect the figure.

rhombus

right angle   An **angle** equal to 90 **degrees**, or $\frac{\pi}{2}$ **radians**. It is a **quarter** of a full turn.

right angle   Common system of making a right angle.

roman numerals A system of writing the **natural numbers**, used by the Romans:

$$I=1 \quad V=5 \quad X=10 \quad L=50$$
$$C=100 \quad D=500 \quad M=1000$$

All other numbers, are made using combinations of the letters above:

4=IV	(1 before 5)
6=VI	(1 after 5)
27=XXVII	(7 after 20)
90=XC	(10 before 100)
1982=MCMLXXXII	

root In an **equation** a **value** which when substituted into the equation in place of the unknown, will make the equation true. For example:

the roots of $x^2-7x+12=0$ are 3 and 4
since $3^2-7\times3+12=0$
and $4^2-7\times4+12=0$
2 is not a root because $2^2-7\times2+12\neq0$.

A number is said to be a root of another if by taking **powers** of the number it is possible to make the other.
(a) 5 is the square root of 25 since 5×5 or $5^2=25$.
(b) 2 is the cube root of 8 since 2×2 or $2^3=8$.
(c) 3 is the fifth root of 243 since $3^5=243$.

rotation A geometrical **transformation** where

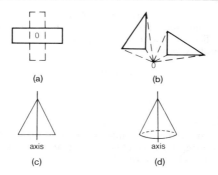

rotation Both the rectangle (a) and the triangle (b) have rotated 90° about the point 0. Rotation of the triangle (c) about its axis gives a cone.

every **point** turns through the same **angle**. In a **plane**, rotation is about a single point called the **centre** of rotation. The centre of rotation is the only point which does not change its position after the rotation.

To distinguish between **clockwise** and anti-clockwise rotations, + and − signs are used, − for clockwise and + for anticlockwise. For example, +60° means a rotation of 60° anticlockwise,

$-90°$ means a rotation of 90° clockwise.

Note that $+90°=-270°$ and $-100°=+260°$, etc.

Rotations with centre (0,0) can be expressed in **matrix** form. For example:

$$+90° \text{ about } (0,0)=\begin{pmatrix} 0 & -1 \\ 1 & 0 \end{pmatrix}$$

$$-90° \text{ about } (0,0)=\begin{pmatrix} 0 & 1 \\ -1 & 0 \end{pmatrix}$$

An object can be rotated in space about an **axis** of rotation. For example, a **cone** is obtained by rotating an **isosceles** triangle about its axis of **symmetry**. See **revolve**.

rounding error   When a number is rounded off to a certain number of **digits** to obtain an **approximation**, the difference between the number and the approximation is called the *rounding error*.

For example, when 1.875 is rounded off to 1.88 the rounding error is $1.8-1.875=0.005$.

Sometimes even very small rounding errors can have a significant effect. This usually occurs in very long calculations with very many operations (**additions**, **subtractions**, **multiplications** or **divisions**).

Some calculators are also prone to rounding errors. Some of them give the solution to $(7\div3)\times3$ as 6.9999999. This occurs because the calculator only works to a set number of **decimal** places. It works out 7÷3 as 2.3333333 instead of 2·3, so there is a rounding error of 0.00000003. Then when it multiplies by 3 it obtains 6.9999999. This makes it important to recognize that 6.9999999 is almost exactly 7. See **rounding off**.

rounding off A way of **approximating** a number to a number with fewer non-zero digits. For example:
(a) Round off £16.25 to the nearest £1=£16.
(b) Round off 275 m to the nearest 100 m=300 m.
(c) Round of 25471 to the nearest 1000=25000.

When the first digit to be ignored is a 0, 1, 2, 3, 4, we 'round down'; when it is a 5, 6, 7, 8, 9 we 'round up'. For example:

```
27381  to the nearest  100 is
27400                                (round up)
42.73  to the nearest 1 is 43        (round up)
6.1238 to the nearest 1/10 is 6.1    (round down)
865    to the nearest 10 is 870      (round up)
(The first digit being ignored is underlined.)
```

See **rounding error**.

row A list of numbers or letters written horizontally. For example:

1 2 3 4 5 6

(2, 5)   **coordinates** are written as a row.

$\begin{pmatrix} 1 & 2 & 5 \\ 3 & 5 & 6 \end{pmatrix}$ this **matrix** has two rows.

(1 7 6 4 2)   a matrix containing only one row is
                               called a row matrix.

ruler A straight **edge** marked in **linear** units
and used for measuring distances. See **scale**.

S

sample A finite proportion of a **population** (which may be **infinite**). For example:
(a) A sample from the set of integers could be 1, 3, 5, 7, 9, 11 or 1, 4, 9, 16, 25 or -1, -2, -3, -8, etc.
(b) A sample from the letters in the alphabet could be {vowels} or {consonants}.

satisfy 1. To fulfil the conditions of. For example, 120° satisfies the requirements of being an **obtuse** angle.
2. A **value** or **set** of values is said to satisfy an **equation** if, when they are **substituted** into the equation, they make the equation balance. For example:

$x=3$ satisfies $x^2+5=14$ since $3^2+5=14$ is true.

$x=2$, $y=3$ satisfy the equations $x+y=5$ and $2x-y=1$ since $2+3=5$ and $2\times2-3=1$ are both true.

scalar A quantity that has size only but not direction (as opposed to a **vector** which has both size and direction). For example, time, distance, mass, **volume** are all scalars.

scale 1. A series of marks on a **ruler**, measuring cylinder, etc., used as an aid in measuring distances, volumes etc.

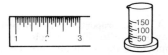

scale   Examples of scales.

2. A scale drawing of an object has all measurements in the same **ratio** with the corresponding measurements of the original.

A map scale of 1:50 000 means that 1 cm on the map represents 50 000 cm − (0·5 km) on the ground.

scale   A scale drawing.

scale factor When a shape is **enlarged** or reduced, the scale factor is defined as:

distance between any two points on the image

the corresponding distance on the object

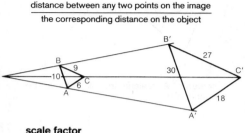

scale factor

$\dfrac{A'B'}{AB} = \dfrac{A'C'}{AC} = \dfrac{B'C'}{BC} = \dfrac{3}{1}$ so the scale factor is 3.

If the areas of the shapes illustrated were compared, the enlarged shape would have an area nine times larger than the original shape. We say that the **area** scale factor is nine.

If a solid shape is enlarged, for example, a **cuboid**, $2 \times 3 \times 4$ becomes a cuboid $4 \times 6 \times 8$, we notice that all linear measurements are doubled, all areas are multiplied by four and the volume is eight times larger. We say:

> linear scale factor $= 2$
> area scale factor $= 4$ (note $4 = 2^2$)
> volume scale factor $= 8$ (note $8 = 2^3$)

scalene A **triangle** is one where no two sides or **angles** are **equal**.

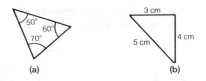

scalene A scalene triangle (a) may be right-angled (b).

scatter diagram A diagram (see opposite page) showing the joint values of two variables.

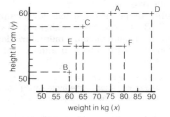

(a)

	Weight (kg)	Height (cm)
A	75	60
B	60	51
C	65	58
D	90	60
E	62	55
F	80	55

(b)

scatter diagram The diagram (a) shows the values in the table (b).

secant A **trigonometric** function. In a **right-angled** triangle the secant of an **angle** is the **ratio**:

$$\frac{\text{hypotenuse}}{\text{adjacent side}}$$

It is usually written sec α. The value of α can be determined by examining a table of values for the secant function.

secant Sec $\alpha = \dfrac{AC}{AB} = \dfrac{10}{8} = 1.25.\,\text{Sec}$

$\alpha = \dfrac{1}{\cos \alpha}$. Using a table of values, $\alpha = 36.9°$.

second 1. A **unit** of time. There are 60 seconds in 1 **mintue**.

2. A unit of circular measure. Each one **degree** of **angle** can be subdivided into 60 minutes ($1° = 60'$) and each minute can be subdivided into 60 seconds ($1' = 60''$).

section See **cross-section**.

sector A part of a **circle** enclosed between two radii.
(a) The area of a sector of $\theta = \theta/360 \times \pi R^2$.
(b) The arc length of a sector of $\theta°$ is $\theta/360 \times 2\pi R$ where R is the **radius** of the circle.
See **segment, quadrant**.

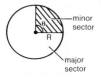

sector Minor and major sectors.

segment   A part of a circle cut off by a **chord**. When the chord goes through the **centre** of the circle the two segments are called **semi**-circles.

major segment

minor segment

segment   Major and minor segments.

self-inverse   A **member** of a **set** is said to be self-inverse if it combines with itself to give the **identity** element of the set under a given **operation**. For example:

the **matrix** $\begin{pmatrix} -1 & 0 \\ 0 & -1 \end{pmatrix}$ is self-inverse under matrix multiplication since

$$\begin{pmatrix} -1 & 0 \\ 0 & -1 \end{pmatrix} \times \begin{pmatrix} -1 & 0 \\ 0 & -1 \end{pmatrix} = \begin{pmatrix} 1 & 0 \\ 0 & 1 \end{pmatrix}.$$

For example, the **element** 2 is self-inverse under addition **modulo** 4 since $2+2=0$.

semi-circle   Half a **circle** formed when a **chord** passes through the centre of a circle.

semi-circle Area=$\frac{1}{2}\pi R^2$. Length of semi-circle **arc**=πR.

semi-interquartile range Of a set of numbers, defined as: $\frac{1}{2}$ (upper quartile−lower quartile). For example, 1, 3, 7, 8, 9, 10, 12, 13, 17, 20, 27.

$$\left.\begin{array}{l}\text{Upper quartile}=17\\\text{Lower quartile}=7\end{array}\right\}\begin{array}{l}\text{semi-interquartile range}\\=\frac{1}{2}(17-7)=5.\end{array}$$

sense A word used in connection with **transformations**. If the shape transformed is 'flipped over' by the transformation we say a change of sense has occurred. **Rotations**, **translations**, **enlargements** and **shears** do not change the sense of a shape, but **reflections** always do.

sequence A set of quantities, often numbers, that are ordered a_1, a_2, a_3 ... so that each **member** of the sequence corresponds to a particular **natural** number. Sequences are sometimes defined by a **formula**, for example, $a_n=2n^2-1$ yields the sequence 1, 7, 17, ... Sequences can also be defined inductively by relating each term to its predecessors, for example, $a_n=3a_{n-1}+1$ and $a_1=1$ yields the sequence 1, 4, 13, 40 . . .

Sequences can have a finite or **infinite** number of terms.

series The **sum** of a finite or **infinite** number of terms, $a_1+a_2+a_3+$ . . .

The study of a series and associated **properties** relating to **sums** and **limits** is an important branch of pure mathematics.

The sum to n terms of the series of terms in an **arithmetic progression** $a+(a+d)+(a+2d)+(a+3d)$ . . . is:

$$\frac{n}{2}(2a+[n-1]d).$$

The sum to n terms of the series of terms in a **geometric progression** $a+ar+ar^2+ar^3$ . . . is:

$$\frac{a(1-r^n)}{1-r}$$

set A collection of distinct objects or things. A set comprises **members** or **elements**, of which there may be a finite or **infinite** number.

The symbol { } means 'the set of', and the symbol ε means 'is a member of'.

Sets may be defined by listing or describing elements, and are often denoted by a capital letter. For example:

E = {even numbers} and 4εE
V = {a, e, i, o, u} and iεV.

sexagesimal system A system related to a
base of 60. For example, the common system of
degree measure for angles is a sexagesimal
system.

360 degrees = 1 revolution
60 **minutes** = 1 degree
60 **seconds** = 1 minute

shear A **transformation** in which a line (in
two dimensions) or a **plane** (in three dimensions)
remains fixed whilst all other **points** move,
parallel to the fixed line or plane, a distance
proportional to their distance from the line or
plane.

shear All points move parallel to a fixed
point.

The **area** of a closed figure in a plane remains
invariant under a shear. Likewise the **volume**

of a **solid** remains invariant under a shear.

Shears with invariant lines or planes through the origin of **coordinates** can be represented by a **matrix**. For example:

$$\begin{pmatrix} 1 & 0 \\ 3 & 1 \end{pmatrix}$$

represents a shear in the $x - y$ plane with the y-axis being held fixed.

Various geometrical and spatial properties of figures can be explored through the use of the shearing transformation.

shear The area of a **parallelogram** is equal to the area of a rectangle by a shear.

side In a **polygon** one of the line **segments** forming the **boundary** of the figure. For example, a **quadrilateral** has four sides.

sieve of Eratosthenes An ancient process for finding **prime numbers** by striking out from the list of **natural** numbers firstly the multiples of

2(4, 6, 8 . . .), then 3, then 5, etc., thus leaving the prime numbers.

(Named after the Greek mathematician Eratosthenes (*c.* 276–194 BC).)

sigma A letter of the Greek alphabet. Upper case Σ and lower case σ.

The former of these two **symbols** is often used to signify a summation. For example:

$$\sum_1^5 a_i = a_1 + a_2 + a_3 + a_4 + a_5$$

$$\sum_1^4 i^2 = 1^2 + 2^2 + 3^2 + 4^2 = 30$$

The lower case symbol σ is often used to signify the **standard deviation** of a **distribution** in **statistics**.

sign Notation to signify whether a quantity is **positive** or **negative**. Positive sign +, negative sign −.

Also used in relation to the **symbols** representing various mathematical **operations** and relations. For example, multiplication sign ×, equals sign =, square root sign √.

significant 1. In **statistics**, a difference or deviation between observations and a proposed theory is called significant if it is unlikely to be the result of chance fluctuations.
2. The significance of a particular **digit** in a number is concerned with its relative size and importance in the number, and is dependent on the place value of the **digit**. For example, in the number 39.08, 3 is the most significant digit, whilst 8 is the least significant.

Measurements are often **rounded off** to prescribed numbers of significant figures. For example, 27.058 can be rounded off thus:

> 30 to 1 significant figure
> 27 to 2 significant figures
> 27.1 to 3 significant figures, etc.

similar Alike. Objects, **plane** or **solid** figures, are similar if they have the same shape.

(a)

similar (a) Objects of the same shape, but not necessarily the same size. (b, overleaf) In an enlargement the shape and its image are similar.

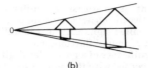

(b)

When a shape is subject to an **enlargement**, the shape and its **image** are similar.

The corresponding **angles** contained in similar figures are equal, and the **ratio** of the lengths of corresponding sides of similar figures is **constant**, as the diagram below shows.

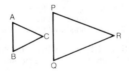

similar ∠BAC= ∠QPR,
∠ACB= ∠PRQ, ∠ABC= ∠PQR
AB:PQ=BC:QR=CA:RP.

simple closed curve A **curve** that can be drawn with a single sweep of a pencil, with the end of the curve joined to the beginning, and such that the curve does not cross itself at any point. See **closed curve**.

simple closed curves

simple harmonic motion (SHM)   The motion of a body that oscillates about a fixed **point**, its **acceleration** being directed towards the fixed point, and proportional to its distance from the point.

The time for a whole oscillation is called the **period** of the motion. The extreme distance from the **centre** is called the **amplitude**.

Motions of this type can be modelled by the projection onto a diameter of a particle moving with constant angular velocity ω round a circle.

$$\text{Acceleration} = -\omega^2 x$$
$$\text{Distance from centre } x = a \cos(\omega t + \alpha)$$
$$\text{Period} = \frac{2\pi}{\omega}$$

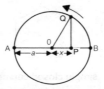

simple harmonic motion As Q moves round the circle, P moves along AB in SHM about 0.

simplify To contract an expression by the use of algebraic or arithmetical techniques. For example, $\frac{6}{9}=\frac{2}{3}$ or $3x+2x+4x=9x$.

Simpson's rule A method for finding the approximate area under a **curve**.

The method derives from dividing the interval over which the **integral** is defined into strips of equal width bounded above by **quadratic** curves, which are taken to approximate closely to the function being considered.

The value of $\int_a^b f(x)\ dx$ given by Simpson's rule with $2n$ strips of width h is:

$$\frac{1}{3}h[y_0+y_{2n})+4(y_1+y_3+\ldots)+2(y_2+y_4+\ldots)]$$

For example, using the rule with four strips of unit width:

$$\int_1^5 x^3 dx \simeq \frac{1}{3} \times 1 \times [(1+125)+4(8+64)+2(27)]$$

$$\simeq \frac{468}{3}$$

$$\simeq 156$$

(Named after the English mathematician Thomas Simpson (1710–61).)

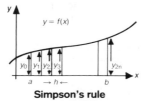

Simpson's rule

simultaneous equations A system of several equations with several unknowns (often **linear**). Examples:

$$\left.\begin{array}{l} 3x+y=17 \\ 2x-2y=12 \end{array}\right\} \quad \text{or} \quad \left.\begin{array}{l} x^2y+y^2=26 \\ x+y=6 \end{array}\right\}$$

$$\left.\begin{array}{l} 3p-2q+5r=6 \\ \text{or } 2p+3q+4r=7 \\ 4p+3q-2r=8 \end{array}\right\}$$

Linear simultaneous equations may be solved in variety of ways:

(a) substitution;
(b) graphical;
(c) addition/subtraction;
(d) matrices;
explained in detail below.

(a) Substitution

$$x+3y=11$$
$$5x-2y=4$$

From the first equation $x=11-3y$.
Substituting in the second equation gives:

$$5(11-3y)-2y=4$$
$$55-15y-2y=4$$
$$17y=51$$
$$y=3$$
$$\text{So, } x=11-(3\times3)$$
$$x=2$$

(b) Graphical

$$x+3y=11$$
$$5x-2y=4$$

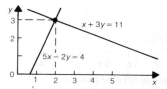

simultaneous equations The point of **intersection** of the lines (2, 3) gives the solution of the equations.

(c) Addition/subtraction

$$x+3y=11$$
$$5x-2y=4$$

multiply the first equation by 5:

$$5x+15y=55$$

then subtract the second equation:

$$17y=51$$
$$y=3$$

now substitute to find x:

$$x+3\times3=11$$
$$x=2$$

(d) Matrices

The equations:

$$x + 3y = 11$$
$$5x - 2y = 4$$

can be written:

$$\begin{pmatrix} 1 & 3 \\ 5 & -2 \end{pmatrix} \begin{pmatrix} x \\ y \end{pmatrix} = \begin{pmatrix} 11 \\ 4 \end{pmatrix}$$

Applying the inverse **matrix**:

$$\frac{1}{17} \begin{pmatrix} 2 & 3 \\ 5 & -1 \end{pmatrix}$$

to each side of this matrix equation, we have:

$$\begin{pmatrix} 1 & 0 \\ 0 & 1 \end{pmatrix} \begin{pmatrix} x \\ y \end{pmatrix} = \frac{1}{17} \begin{pmatrix} 2 & 3 \\ 5 & -1 \end{pmatrix} \begin{pmatrix} 11 \\ 4 \end{pmatrix}$$

$$\Rightarrow \begin{pmatrix} x \\ y \end{pmatrix} = \begin{pmatrix} 2 \\ 3 \end{pmatrix}$$

Simultaneous equations arise from a variety of different physical situations, and the study of their solution forms an important branch of algebra. The idea of a matrix was developed through the study of equations of this type.

sine A **trigonometric** function. In a **right-angled** triangle the sine of an angle is the **ratio**:

$$\frac{\text{opposite side}}{\text{hypotenuse}}$$

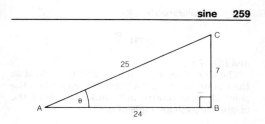

sine   In $\triangle ABC$ $\sin\theta = \dfrac{CB}{AC} = \dfrac{7}{25} = 0.28.$

In $\triangle ABC$ above the value of θ can now be determined by examining a table of values of the sine function: when $\sin\theta = 0.28$, $\theta = 16.3°$.

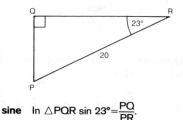

sine   In $\triangle PQR$ $\sin 23° = \dfrac{PQ}{PR}.$

For $\triangle PQR$, from the table of values of the sine function $\sin 23° = 0.391$. Hence:

$$0.391 = \frac{PQ}{20}$$

and PQ = 7.82.

The sine of an angle θ may also be defined as the *y* **coordinate** of the point obtained when the point (1, 0) is rotated anti-clockwise, about the origin, through an angle of θ.

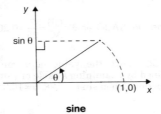

sine

singular matrix A square **matrix** that has no inverse. For example:

$$\begin{pmatrix} 3 & 6 \\ 2 & 4 \end{pmatrix} \quad \text{or} \quad \begin{pmatrix} 2 & 1 & 3 \\ 1 & 4 & 5 \\ 4 & 9 & 13 \end{pmatrix} \quad \text{or} \quad \begin{pmatrix} 1 & 2 & 3 \\ 0 & 0 & 0 \\ 4 & 5 & 6 \end{pmatrix}$$

The **determinant** of a singular matrix is **zero**, and there are simple **linear** relationships between the **rows** of the matrix and between the **columns** of the matrix.

Singular matrices are associated with **sets** of **linear** equations without unique solutions. For example:

matrix $\begin{pmatrix} 3 & 6 \\ 2 & 4 \end{pmatrix}$

$3x+6y=9$
$2x+4y=6$
system with
many solutions

$3x+6y=1$
$2x+4y=7$
system with
no solutions

skew lines Non-**parallel**, non-**intersecting** lines.

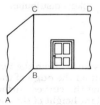

skew lines The line AB marking the boundary between floor and wall is skew to the line CD between wall and ceiling.

skew distribution Data which when represented graphically produce a non-**symmetrical** profile.

The **bar chart** illustrated represents the **frequencies** of various numbers of goals scored by soccer teams in a league. The distribution is skew, the frequencies do not group around a central **average**.

skew distribution

slant The distance *l* from the **vertex** of a right-**cone** to a point on the **edge** of the circular base, measured along the curved surface of the **cone**, is called the slant height of the cone.

This measurement is used in calculating the area of the curved surface of a right cone. Area=πrl.

slant The slant height, *l*, of a cone.

slide rule A calculating aid looking similar to a simple rule, but containing a sliding **logarithmic** scale, which can be used to multiply, divide, extract **roots**, etc.

Slide rules were extensively used by engineers, draughtsmen and others involved in doing large numbers of calculations. It has been largely replaced by the electronic calculator as a practical calculating aid.

slide rule

small circle A circle drawn on the surface of a **sphere**, whose **centre** does not coincide with the centre of the sphere.

The term is used most commonly with reference to the geometry of the earth. Lines of **latitude** are small circles (apart from the **equator**, which is a **great circle**) whose centres lie along a common **diameter** of the earth.

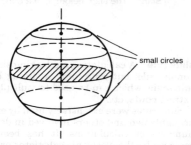

small circles

small circles The centres do not coincide with the centre of the sphere.

solid of revolution The **solid** formed by rotating a plane surface area through 360° about an axis.

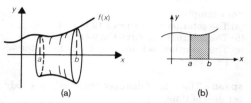

(a)   (b)

solid of revolution (a) The solid formed and (b) the flat area which sweeps out the solid.

The diagram shows the solid formed by rotating the area under the curve $y=f(x)$ between $x=a$ and $x=b$ about the x-axis. The flat area which sweeps out the solid is:

$$\int_a^b y \, dx$$

The surface area of the solid of revolution is:

$$\int_a^b 2\pi y \sqrt{1+\left(\frac{dy}{dx}\right)^2} \, dx$$

The volume of the solid of revolution is:

$$\int_a^b \pi y^2 \, dx$$

solution The process and result of solving a mathematical problem often in **equation** form.

For example:
(a) The solution to $3x+4=x+12$ is $x=4$.
(b) The solution to $dy/dx=x$ is $y=x^2/2+A$.
(c) The solution set to $x^2 \leq 4$ is the interval $[-2, 2]$.

speed   The rate of distance travelled by a body per unit of time.

Average speeds are calculated over a time interval by:

$$\frac{\text{distance travelled}}{\text{time taken}}$$

For example, a train travels 200 km in $2\frac{1}{2}$ hours. Average speed:

$$\frac{200}{2\frac{1}{2}}=80 \text{ km/hour.}$$

The instantaneous speed of a body at a particular time is the **limit** of the average speeds over a **sequence** of time intervals that approach zero.

In the graph, the average speed over the second and third seconds is measured by the **gradient** of the **chord** AC. Average speed:

$$\frac{45-5}{3-1}=20 \text{ m/s}$$

The instantaneous speed after two seconds is measured by the gradient of the **tangent** at B. Speed:

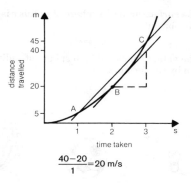

$$\frac{40-20}{1}=20 \text{ m/s}$$

speed The distance travelled by a body in free fall.

sphere A circular **solid** formed by the **set** of **points** in space all equidistant from a given fixed point.

The fixed point is called the **centre** of the sphere, and the common distance of its surface points from the centre is called the **radius**.

The **Cartesian** equation of a sphere with centre at (a, b, c) and radius r is:

$$(x-a)^2+(y-b)^2+(z-c)^2=r^2$$

The surface area of a sphere of radius r is given by the formula:

$$A = 4\pi r^2$$

The volume enclosed by a sphere of radius r is given by the formula:

$$V = \frac{4}{3}\pi r^3$$

sphere A football, a soap bubble and a globe are all examples of roughly spherical shapes.

spiral A general term given to plane curves or curved surfaces with coils which resemble characteristic patterns found in the natural world.

The logarithmic or equiangular spiral can be constructed by successive dissections of a rectangle into squares, and the **inscribing** of **quadrants** of circles.

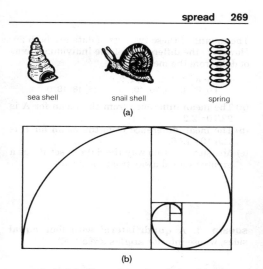

sea shell   snail shell   spring

(a)

(b)

spiral (a) Examples of spirals (b) a logarithmic spiral.

spread A term used to denote the dispersion of **data** from a measure of **average**. Two commonly used measures of spread are **interquartile range** and **standard deviation**. Examples:

Data A   19, 21, 15, 25, 20, 22, 18, 20, 17, 23
Data B   20, 0, 40, 30, 10, 5, 35, 2, 38, 20

The **means** of these two sets of data are both 20. However, the differences of the individual items of data from the means are:

    For A   1, 1, 5, 5, 0, 2, 2, 0, 3, 3
    For B   0, 20, 20, 10, 10, 15, 15, 18, 18, 0

(a) The mean difference from the mean for A is 22/10=2.2.
(b) The mean difference from the mean for B is 126/10=12.6.
(c) Measured in this way the data in set B has a greater spread away from the mean.

square 1. A **quadrilateral** with four **equal** sides, whose interior **angles** are all 90°.

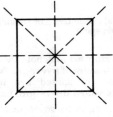

square

A square is **symmetrical** about its **diagonals** and the **perpendicular bisectors** of its sides, which are all **concurrent**. A square has rotational **symmetry** of **order** 4.

2. To square a quantity is to multiply that quantity by itself. For example, 5 squared (written 5^2)$=5\times5=25$.

3. Square **centimetre** (cm^2), square **metre** (m^2), square **kilometre** (km^2) are all **units** used in measuring **area**. A surface with area 5 cm^2, has the same area as five squares of side 1 cm.

square root The number x is a square **root** of the number y ($x=\sqrt{y}$) if y is the square of x ($y=x^2$).

9 is a square root of 81 (written $\sqrt{81}$) since $81=9\times9$.

Positive **real numbers** have two square roots. For example, the square roots of 6.25 are $+\sqrt{6.25}=2.5$ and $-\sqrt{6.25}=2.5$.

standard deviation A commonly used measure of the **spread** of individual items for **data** from the **mean** of the **set** of items. For example, the heights in centimetres of ten small plants are:

6.3, 5.9, 5.2, 7.1, 6.6, 6.8, 7.4, 5.3, 5.8, 5.6

The mean height is 62/10=6.2 cm.

The deviations of each of the individual heights from the mean are:

+0.1, −0.3, −1.0, +0.9, +0.4, +0.6, +1.2, −0.9, −0.4, −0.6

The **squares** of these deviations are:

0.01, 0.09, 1.0, 0.81, 0.16, 0.36, 1.44, 0.81, 0.16, 0.36

The mean of these squares of the deviations is the **variance**:

$$\frac{5.2}{10} = 0.52.$$

The **square root** of this mean is 0.72.

This quantity, the root, mean, square, deviation from the mean, is called the **standard deviation**.

The **formula** for calculating standard deviation is:

$$\sqrt{\frac{\sum_{1}^{n} (x_i - \bar{x})^2}{n}}$$

for n items of data $x_1, x_2 \ldots$ with a mean of $\bar{x}$.

standard index form A method of expressing numbers, popular in science and engineering, in which positive numbers are written in the form:

$$A \times 10^n$$

where $1 \leq A < 10$ and n is an **integer**. For example:

$$734.6 = 7.346 \times 10^2$$
$$-0.0063 = -6.3 \times 10^{-3}$$

This notation is particularly useful for dealing with very large or small numbers.

stationary point A **function** $y=f(x)$ has a stationary point at (a, b) if the **derivative** of $f(x)$ is zero at $x=a$, i.e. $f'(a)=0$.

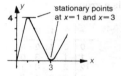

stationary points at $x=1$ and $x=3$

stationary points $f(x)=x^3-6x^2+9x$

The **gradient** of the **graph** of a function is **zero** at a stationary point and the **tangent** is **parallel** to the x-axis. Three types can be identified:

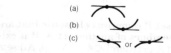

(a)
(b)
(c)   or

stationary points (a) Maximums, (b) minimums, (c) stationary inflections.

statistics The study of methods of analysing large quantities of **data**.

The measures of the various properties of **sets** of data, for example, **means**, **deviations**, etc., are also known as statistics.

subgroup A **group** H whose **elements** form a **subset** of a group G, and that shares the same **operation** with G, is called a subgroup of G.

For example, the set {0, 1, 2, 3} under addition **modulo** 4 is a group:

+	0	1	2	3
0	0	1	2	3
1	1	2	3	0
2	2	3	0	1
3	3	0	1	2

The subset {0, 2} under the same operation forms a subgroup:

+	0	2
0	0	2
2	2	0

subset If a **set** B comprises **elements** that are themselves **members** of another set A, B is called a subset of A (written B⊂A or A⊃B). All sets are subsets of a **universal set** $\mathscr{E}$. The **empty set** ∅ is a subset of every set.

If A contains **elements** not in B then B is a
proper subset of A.

subset B⊂A.

substitution The replacement of one expression by another in order to **simplify**. The replacement of a **variable** by a number, for example, substituting $x=3$ into the **equation** $y=2x+1$ gives $y=2\times3+1=7$.

It is particularly used as a method of simplifying certain **integrals**. For example:

$$\ln \int \frac{1}{\sqrt{1-x^2}}\, dx \quad \text{substitute } x=\sin\theta$$

$$= \int \frac{1}{\sqrt{1-\sin^2\theta}}\, \cos\theta\, d\theta$$

$$= \int 1.d\theta = \theta + K$$

$$= \sin^{-1}x + K$$

subtend angle An **angle** is subtended by an

arc or line when the angle 'stands upon' the arc or line.

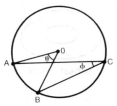

subtend The arc AB subtends the angle φ at the **circumference** and the angle θ at the **centre**.

subtraction One of the four basic operations of **arithmetic**. It is related to situations where one quantity is removed from another or when the difference between two quantities is to be calculated. Subtraction is the **inverse** operation to **addition**.

sufficient condition Considering the statement 'if x is 3 and y is 4; then x+y is 7', of the two conditions:
(1) 'x is 3 and y is 4'
(2) 'x+y is 7'
(1) is a sufficient condition for (2), i.e. if (1) is

true, then (2) follows. However, (2) is not a sufficient condition for (1), since if (2) is true (1) need not follow.

sum The result of adding two or more quantities.

supplementary angles Two **angles** whose **sum** is 180°.

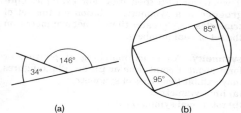

(a)   (b)

supplementary angles (a) Supplementary angles standing on a straight line; (b) supplementary angles in a **cyclic quadrilateral**.

surd An expression involving unresolved **roots** of numbers. For example:

$$\sqrt{2}, \sqrt{5}+\sqrt{7}, \sqrt[3]{8}, \sqrt{3}-\sqrt{2}, \text{etc.}$$

Some expressions involving surds can be simplified. For example:

$$\sqrt{8}=\sqrt{4\times2}=\sqrt{4}\times\sqrt{2}=2\sqrt{2}$$

symbol A sign to denote a **quantity**, **operation**, **relation** or other mathematical entity. Examples:

$$+ \quad \{ \; \} \quad \vee \quad \int \quad = \quad x$$

symmetric A **relation** R on a set is symmetric if whenever aRb then bRa. For example, **congruence** is a symmetric relation on the set of **triangles**, but 'greater than' is not symmetric on the set of **real** numbers.

symmetry An exact matching of position or form about a point, line or **plane**. Plane figures may exhibit two types of symmetry:
(a) line symmetry;
(b) rotational symmetry.
(a) Line symmetry
The figure (a) opposite is equally distributed about the two dotted lines of symmetry. It is unchanged by a reflection in one of these lines. It has line symmetry of **order** 2.
(b) Rotational symmetry
The figure (b) opposite maps onto itself under **rotations** of 120°, 240° and 360° about the centre **point**. It has rotational symmetry of order 3.

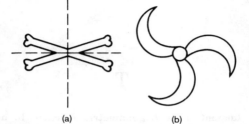

symmetry (a) Line symmetry and (b) rotational symmetry.

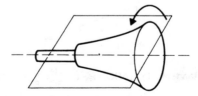

symmetry Solid shapes may have symmetry about a plane, or about an **axis** of rotation.

T

tangent 1. A **trigonometric** function. In a **right-angled** triangle, the tangent of an angle is the ratio:

$$\frac{\text{opposite side}}{\text{adjacent side}}$$

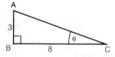

tangent In $\triangle ABC$, $\tan \theta = \dfrac{AB}{BC} = \dfrac{3}{8} = 0.375$.

From a table of values of the tangent function the value of θ in $\triangle ABC$ above can be found: $\theta = 20.6°$.

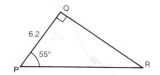

tangent   In △PQR, $\tan 55° = \dfrac{QR}{6.2}$

For △PQR, from a table of values of the tangent function $\tan 55° = 1.43$.

So $\dfrac{QR}{6.2} = 1.43$ and $QR = 6.2 \times 1.43 = 8.87$

The tangent of an angle θ may also be defined by

$$\tan \theta = \frac{\sin \theta}{\cos \theta}$$

2. A tangent to a **curve** at a **point** is a straight line that touches the curve at that point, and has the same **gradient** as the curve at that point. If P has **coordinates** (a, b) and the gradient of the curve at P is m then the equation of the tangent at P is $y - b = m(x - a)$. See diagram overleaf.

terminating decimal   A **decimal** that has a

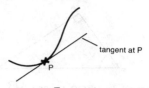

tangent Tangent to a curve.

finite number of decimal **digits**. For example, 0.41, 1.825, 0.1629385794.

All terminating decimals are equivalent to **fractions** having **denominators** whose **prime factors** are from the set $\{2, 5\}$. **For example:**

$$\frac{3}{40} = 0.075 \quad \frac{7}{125} = 0.056 \quad \frac{13}{200} = 0.065 \text{ etc.}$$

See **recurring decimal**.

tessellation A regular pattern of tiles covering a surface is called a tessellation.

tessellation

tetrahedron A **polyhedron** with four triangular faces. A triangular-based **pyramid**.

tetrahedron

theorem A significant general conclusion obtained by logical deduction from certain assumptions called **hypotheses**. For example, **Pythagoras theorem**. See **proof**.

tonne A tonne is a unit of mass equivalent to 1000 **kilograms**.

topology The study of certain geometrical properties of figures that remain unchanged under particular types of **transformations**.

topology

For simple plane **networks** properties such as the number and types of **nodes**, the number of **arcs**, etc., are topological properties which remain unchanged under the transformations considered.

torus   A doughnut-shaped surface.

It is the result of rotating in space a **circle** of **radius** a, about an **axis** in the **plane** of the circle, at a distance b from the *centre* of the circle.

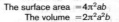

The surface area $= 4\pi^2ab$
The volume $= 2\pi^2a^2b$

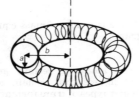

torus

trajectory   The path of a moving body or **projectile**. It is often used in the context of bodies moving under the influence of the earth's (or other large object's) gravitational field.

trajectory The path of a moving cricket ball.

transcendental numbers Irrational numbers such as π (**pi**) and **e**, which cannot be obtained as the result of solving a **polynomial** equation with **rational coefficients**.

Other numbers, including irrationals, which can result from solving such **equations** are called *algebraic*:

$x^2 = 2$ has $\sqrt{2}$ as a solution.
Hence $\sqrt{2}$ is algebraic.

transformation A change in spatial or algebraic work involving the **mapping** of one figure or space to another, or one expression to another, often by use of a **substitution**.

The study of various types of transformations, for example, **rotations**, **translations**, is central to work in both **algebra** and **geometry**.

transitive relation   A **relation** R between **elements** of a **set** whereby it follows that when any three elements x, y and z are related xRy and yRz, than xRz.

For example, the relation $>$('is greater than') defined on the set of real numbers is transitive: $7>6$ and $6>4$ implies that $7>4$.

The relation $\uparrow$ ('is half of') defined on the set of real numbers is not transitive: $3 \uparrow 6$ and $6 \uparrow 12$ does not imply $3 \uparrow 12$.

translation   A geometrical **transformation** of the **plane** or of space of the form: $x \rightarrow x+a$, $y \rightarrow y+b$, $z \rightarrow z+c$, etc.

A translation is equivalent to referring to a set of new axes which are **parallel** to the original set. Translations are often written in **vector** form, in this case:

$$\begin{pmatrix} x \\ y \end{pmatrix} \rightarrow \begin{pmatrix} x \\ y \end{pmatrix} + \begin{pmatrix} 7 \\ 1 \end{pmatrix}$$

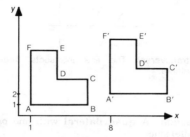

translation The image of a shape after a translation is **congruent** to the original shape.

transpose of a matrix The matrix that results from interchanging the **rows** and **columns** of a given matrix. For example:

$$M = \begin{pmatrix} 3 & 1 \\ 2 & 4 \\ 6 & 5 \end{pmatrix} \quad M' = \begin{pmatrix} 3 & 2 & 6 \\ 1 & 4 & 5 \end{pmatrix}$$

transversal A line that **intersects** two or more other lines (see diagram overleaf).

transversal The line intersects three others.

trapezium A **quadrilateral** with one pair of **parallel** sides.

If the parallel sides are of lengths a and b, and are separated by a distance h, the area of the trapezium is A where:

$$A = \frac{(a+b) \times h}{2}$$

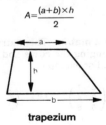

trapezium

tree diagram A diagram in which possible

events are represented by **points** joined by lines.
It is laid out so that a note on each line (or
branch) gives the **probability** of the event at the
end of that line, and so that the rules 'multiply
along the branches' and 'add vertically' apply.

For example, a drawer contains 8 red socks and
4 blue socks. To find the probability of picking a
pair if you take out two socks without looking,
one after the other, the tree diagram shown is
drawn.

See **decision tree**.

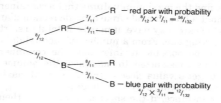

tree diagram

trial   The repetition of a particular experiment
under controlled conditions in a statistical
investigation. See **statistics**.

trial and improvement methods   These
methods do not lead directly to a correct answer
but rather, starting from an **estimate** or an

'answer' known to be near-correct, they lead to increasingly close **approximations** to the correct answer.

For example, to find the side length of a square of **area** 8 cm^2, it is helpful first to observe that $2^2=4<8$ and that $3^2=9>8$. So the side length required must be between 2 and 3. Since $(2.5)^2=6.25<8$, the side length must be more than 2.5. By continuing this method with a calculator, an answer correct to at least three decimal places will be found.

To find the maximum **volume** that a container without a lid and made from a single rectangular sheet of card may have, it may be helpful to start by making one. From a number of such **trials** it should be possible to infer how the container must be constructed to maximize the volume, and to get a value close to the theoretical maximum by working only with diagrams.

For instance, if the sheet is 4 cm×4 cm, then small **squares** of side h are cut from each corner, where h represents the height of the finished container. Then, the volume V is given by:

$$V=(4-2h)\ (4-2h)\times h$$
when $h=1$, $V=2\times2\times1=4$
when $h=\frac{3}{4}$, $V=\frac{5}{2}\times\frac{5}{2}\times\frac{3}{4}=\frac{75}{16}$
when $h=\frac{1}{2}$, $V=3\times3\times\frac{1}{2}=\frac{9}{2}$
when $h=14$, $V=\frac{7}{2}\times\frac{7}{2}\times\frac{1}{4}=\frac{49}{16}$

With heights of $\frac{1}{2}$ and $\frac{3}{4}$ the volume is greater than with heights of 1 and $\frac{1}{4}$, so it is sensible to look for the maximum volume with heights between $\frac{1}{2}$ and $\frac{3}{4}$.

trial and improvement methods
Estimating the volume of a container.

triangle A three-sided **polygon**. Triangles can be classified in a number of ways (see diagram overleaf).
(a) A **scalene** triangle has no two sides equal.
(b) An **isoceles** triangle has two equal sides and thus two equal **angles**.
(c) An **equilateral** triangle has three equal sides and thus three equal angles.
(d) A right-angled triangle has an interior angle equal to a **right angle**.

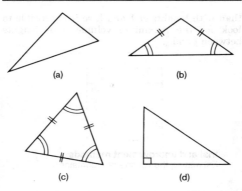

triangles (a) Scalene triangle; (b) isoceles triangle; (c) equilateral triangle; (c) a
right-angled triangle.

trigonometry A branch of mathematics concerned, at its simplest level, with the measurement of **triangles**. Unknown **angles** or lengths
are calculated by using trigonometrical **ratios**
such as **sine**, **cosine** and **tangent**.

Trigonometry also involves the study of the
properties of the various trigonometrical **functions**, and their applications to the solutions of
problems in many branches of mathematics.

trinomial Any **polynomial** with just three terms. For example, x^4+x^2-5 or $3x^2+4x+1$.

truncated Describes a **solid** created from cutting two non-**parallel** planes in the original solid. A truncated **cone** or **prism** is produced from the original solid by removing portions with cutting planes as in the diagrams.

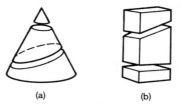

(a)   (b)

truncated The middle sections are (a) a truncated cone and (b) a truncated cuboid.

turning point A **point** on a **graph**, of a **function**, which is a local **maximum** or **minimum** value of the function.

At a turning point the **value** of the **derivative** of the function is **zero**.

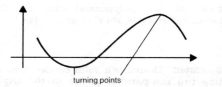

turning points

turning points The maximum or minimum values.

U

unbounded function A **function** is unbounded over an **interval** if a **value** of the function can be found within the **interval** to exceed any given number however big.

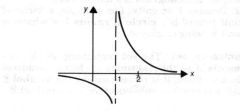

unbounded The function $y=\dfrac{1}{x-1}$ is unbounded in the interval [1, 2].

296   unconditional inequality

unconditional inequality An **inequality** such as $x-1 < x$, which is true for all **values** of x, is called an unconditional inequality.

union A **set** operation. The union of two sets A and B is the set of **elements** contained in A together with the elements contained in B. For example:

If A = {1, 2, 3, 4} and B = {3, 4, 5, 6}
The union of A and B is {1, 2, 3, 4, 5, 6}

The symbol ∪ is used to present the operation.

Hence,   A∪B = {1, 2, 3, 4, 5, 6}.

units 1. The standard measures of mass, length, time. For example, the **metre** is a unit of length, the **kilogram** is a unit of mass.
2. Denotes 1 or unity. For example, a 'circle of unit radius' is a **circle** of **radius** 1 of whatever unit is being employed.

universal set The **set** containing all the **elements** of all the sets which are being considered in a problem, it is represented by the **symbol** $\mathscr{E}$. All sets are thus **subsets** of the universal set $\mathscr{E}$.

upper bound In a **set** of numbers, a number which no **member** of the set exceeds.
   For the set of values, illustrated below, of the function:

$$f(x) = \frac{3x-1}{x}$$

with **domain** the set of positive real numbers, an upper bound is 3.

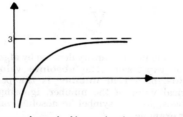

upper bound No number here exeeds 3.

V

value Amount or quantity denoted by algebraic term or expression. The **absolute** value, or **modulus**, of a **real number** is the positive numerical value of the number, ignoring any negative signs. The symbol for absolute value is | |. For example:

$$|-9.3|=9.3 \qquad |7.2|=7.2$$

The value of a **polynomial** for a particular number, is the numerical result of evaluating the polynomial with that number substituted in place of the **variable**. For example, the value of $4x^3-3x^2+7$ at $x=-2$ is -37.

variable A quantity that can take on a range of **values** is called a variable. For a **function** $y=f(x)$, x is called the **independent** variable and can take on any value in the **domain** of the function. y is called the dependent variable and takes on values in the **range** of the function.

For example, for the function $y = \sin x$ the variable x can take on any real number value, whilst y can take on any value in the range $[-1, 1]$.

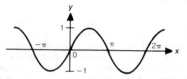

variable y can take on any value in the range $[-1, 1]$.

variance   For n items of **data** x_i with a **mean** of $\bar{x}$, the mean of the squares of the deviations of each x_i from the mean $\bar{x}$.

$$\text{Variance } \sigma^2 = \frac{\displaystyle\sum_{i=1}^{n} (x_i - \bar{x})^2}{n}$$

The **square root** of the variance $\sqrt{\sigma^2}$ is known as the **standard deviation**.

vector   A quantity that has properties of both **magnitude** and direction. Forces, **velocities**, **displacements**, etc., are vectors. The characteristic properties of vectors have been abstracted to produce a theory of vectors, which forms an important branch of pure and applied mathematics.

Vectors are often represented by directed line segments like **AB**, in the diagram, whose length and direction represent magnitude and direction. Vectors are always printed in **bold** type. Vectors add up around vector **polygons** $\mathbf{AB} + \mathbf{BC} + \mathbf{CD} = \mathbf{AD}$. Vectors are often expressed in **component** form relative to some standard basis:

$\mathbf{AB} = \begin{pmatrix} 1 \\ 2 \end{pmatrix}$ relative to the unit vectors along the x- and y-axes.

The magnitude of **AB** is $\sqrt{1^2 + 2^2} = \sqrt{5}$ and its direction is 63.4° to the x-axis $\left(\tan 63.4° = \dfrac{2}{1} \right)$.

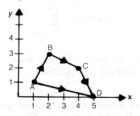

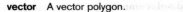

vector A vector polygon.

velocity   A measure of **speed** in a particular direction. It is the rate of change of **displacement** of a body per **unit** time. Velocity is a **vector** quantity.

Venn diagram   A diagram showing the relationships between **sets**, by representing them as **regions** enclosed by **simple closed curves**.

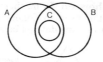

Venn diagram   The relationship between the sets: A={1, 2, 3, 4, 5, 6}, B={4, 5, 6, 7, 8}, C={5, 6}, where C=A∩B.

vertex   A point at which lines or **planes** meet. In a **polygon** the point of **intersection** of two **adjacent sides**. In a **polyhedron** it is the point of intersection of two **edges**.

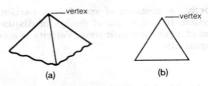

vertex The vertices of (a) a pyramid and (b) a triangle.

vertical The direction **perpendicular** to the **plane** of the **horizon**. In mathematics the term vertical is often used to denote measurements perpendicular to some defined base line.

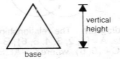

vertical height The vertical height of a triangle.

volume The measure of the amount of space occupied by a **solid**.

The volumes of certain simple solids can be easily calculated:

(a) For a **cuboid**, $V = l \times b \times h$.
(b) For a **prism**, $V = A \times h$.
(c) For a **pyramid** $V = \frac{1}{3} \times A \times h$, where A is the base area.

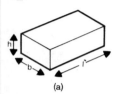

(a)

(b)

(c)

volume The dimensions used for calculating volume of (a) a cuboid, (b) a prism and (c) a pyramid.

vulgar fraction A simple **fraction** with **integer numerator** and **denominator**. For example:

$$\frac{3}{4}, \quad \frac{23}{19}, \quad \frac{721}{859}$$

W

whole numbers 1. The **positive** and **negative** counting numbers with **zero**:

$$\ldots, -5, -4, -3, -2, -1, 0, 1, 2, 3, 4, 5, \ldots$$

Also referred to as **integers**.

2. The positive counting numbers together with zero:

$$0, 1, 2, 3, 4, 5 \ldots.$$

Y

Yard An imperial **unit** of length equal to three feet. There are 1760 yards in one mile. See **foot**.

Z

zero The **cardinal** number associated with the **empty set**. Zero is represented by the numeral 0 and is the **identity** element for **addition**: $x+0=x$ for any **real number** x.

zone A **sphere** bounded by two **parallel** cutting **planes** (see diagram overleaf).

The surface area of a zone of a sphere of radius r, produced by cutting planes distance h apart is:

$$2\pi rh$$

See **truncated**.

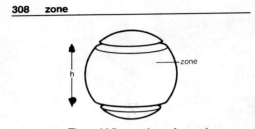

zone The middle section of a sphere created by two parallel cutting planes